STEPS TO COMMON ENTRANCE MATHEMATICS 2

WALTER PHILLIPS

OXFORD
UNIVERSITY PRESS

Contents

Preface

Steps to Common Entrance Mathematics 1, 2 and 3 is a series of three books intended for pupils whose ages range from seven to ten years. Book 2, the second in the series, is for eight- to nine-year-olds. The three books act as a stepping stone to *New Common Entrance Mathematics* written by the same author. The complete set of books cover most of the requirements of the primary syllabuses for Caribbean countries and are designed to reinforce the important concepts taught at primary school.

Each book is arranged in three sections: a section of exercises introducing new material topic by topic, followed by fifteen assessment papers and two tests. Pupils will receive maximum benefit from the books by working through the topics in sequence.

Most chapters finish with a revision exercise designed to help pupils review work covered in that chapter. Included in each book are some interesting cross-number puzzles which allow pupils to have fun solving them.

Walter Phillips
1989

1 Number Concepts

Counting and Writing Numbers

Writing Numbers in Figures and Words

Exercise 1

Write in figures:

1 seventy-three _____

2 sixty-three _____

3 fourteen _____

4 ninety _____

5 thirty-eight _____

6 one hundred and eleven _____

7 forty _____

8 five hundred and thirty-four _____

9 seven hundred and twelve _____

10 eight hundred and forty-five _____

11 seventy-eight _____

12 forty-five _____

13 three hundred and five _____

14 six hundred and twenty   _____

15 four hundred   _____

16 sixty-two   _____

17 two hundred and twenty-eight   _____

18 one hundred and ten   _____

19 nine hundred and five   _____

20 six hundred and twenty-five   _____

Exercise 2

Write in words:

1   43   _____

2   19   _____

3   36   _____

4   25   _____

5   93   _____

6 213   _____

7 408   _____

8   74   _____

9   50   _____

10 510   _____

11 649   _____

12 970   _____

13 308 _____

14 276 _____

15 549 _____

16 619 _____

17 414 _____

18 770 _____

19 271 _____

20 943 _____

Exercise 3

Write in figures:

1 three thousand _____

2 nine thousand _____

3 two thousand, six hundred and thirty-four _____

4 three thousand and seventy-two _____

5 eight thousand and forty-nine _____

6 two thousand and eighty-four _____

7 seven thousand, three hundred and eight _____

8 nine thousand, four hundred and forty _____

9 six thousand and eleven _____

10 four thousand, five hundred and forty-five _____

11 seven thousand and twenty _____

12 four thousand and nine _____

13 one thousand _____

14 one thousand and six _____

15 eight thousand and thirty _____

16 nine thousand and two _____

17 five thousand and five _____

18 one thousand, eight hundred and eleven _____

19 four thousand, one hundred and sixty-five _____

20 five thousand _____

 Exercise 4

Write in words:

1 3726 _____

2 5137 _____

3 6403 _____

4 2708 _____

5 3000 _____

6 5620 _____

7 8064 _____

8 5076 _____

9 2007 _____

10 9001 _____

11 3576 _____

12 4009 _____

13 3074 _____

14 1608 _____

15 9456 _____

16 3470 _____

17 2709 _____

18 8003 _____

19 5600 _____

20 2000 _____

Odd and Even Numbers

> **Example**
>
> Write the missing odd number: 1, *3*, 5, 7

Exercise 5

Write the missing odd number in each row:

1 7, 9, 11, _____

2 15, 17, _____, 21

3 _____, 33, 35, 37

4 41, 43, 45, _____

5 65, _____, 69, 71

6 11, 9, 7, _____, 3

7 27, 25, 23, _____

8 63, 61, _____, 57

9 73, _____, 77, 79

10 _____, 89, 91, 93

Exercise 6

In each row there is one odd number. Circle it.

1 16 18 20 23   6 34 54 68 73

2 29 30 32 38   7 47 50 62 84

3 47 48 62 70   8 12 19 22 40

4 10 12 13 36   9 52 53 54 56

5 26 30 39 48   10 60 61 64 82

Exercise 7

Write the missing even number in each row:

1 6, 8, 10, _____   6 74, 76, 78, _____

2 _____, 16, 18, 20   7 _____, 16, 14, 12

3 _____, 32, 34, 36   8 26, 24, _____, 20

4 50, 52, _____, 56   9 40, _____, 36, 34

5 58, _____, 62, 64   10 98, 96, 94, _____

Exercise 8

In each row there is an even number. Circle it.

1 7 11 14 19   6 15 17 23 28

2 22 23 25 29   7 36 39 41 53

3 27 31 38 41   8 47 51 52 55

4 15 13 6 5   9 70 73 75 79

5 5 8 13 17   10 102 103 105 107

Exercise 9

Divide the following numbers into two groups, odd numbers and even numbers:

7　13　4　23　6　44　12　35　71　25
15　60　91　95　102　107　214　125　147　316

Odd numbers　_____

Even numbers　_____

Place Value

Exercise 10

What is the place value of the digit underlined (hundreds, tens or ones)?

1 4̲7 _____　　**6** 30̲7 _____

2 91̲ _____　　**7** 9̲28 _____

3 6̲6 _____　　**8** 2̲26 _____

4 2̲15 _____　　**9** 44̲4 _____

5 80̲ _____　　**10** 280̲ _____

Exercise 11

What is the place value of the digit underlined (thousands, hundreds, tens or ones)?

1 37̲41 _____

2 8492̲ _____

3 60̲75 _____

4 2̲143 _____

5 7̲006 _____

6 57<u>4</u>0 _____

7 3<u>0</u>00 _____

8 <u>1</u>761 _____

9 643<u>9</u> _____

10 4<u>8</u>3 _____

11 <u>5</u>76 _____

12 4<u>6</u>83 _____

13 8<u>9</u>9 _____

14 <u>1</u>578 _____

15 325<u>0</u> _____

16 In the number 2975, which digit represents the following?

    a   tens   _____   **b**   ones   _____

    c   thousands _____   **d**   hundreds _____

17 Use one of these words: thousands, hundreds, tens, ones, to complete the following. For the number 4076,

    a 7 = 7 _____   **b** 4 = 4 _____

    c 0 = 0 _____   **d** 6 = 6 _____

18 What is the place value of the 7 in each of these numbers?

    a 381<u>7</u> _____   **b** 1<u>7</u>38 _____

    c <u>7</u>318 _____   **d** 31<u>7</u>8 _____

Comparing Numbers

< means less than.

> means greater than.

Exercise 12

Write < or > in the circles to make each statement true:

1 16 ◯ 21

2 19 ◯ 11

3 27 ◯ 38

4 43 ◯ 39

5 50 ◯ 40

6 97 ◯ 63

7 76 ◯ 89

8 86 ◯ 95

9 105 ◯ 76

10 84 ◯ 110

11 116 ◯ 128

12 241 ◯ 263

13 305 ◯ 297

14 375 ◯ 326

15 643 ◯ 751

16 408 ◯ 502

17 376 ◯ 425

18 520 ◯ 486

19 1140 ◯ 980

20 760 ◯ 1048

21 1176 ◯ 848

22 974 ◯ 1002

23 1220 ◯ 996

24 1327 ◯ 1468

25 1534 ◯ 1383

26 2741 ◯ 3406

27 2005 ◯ 3004

28 4126 ◯ 5179

29 3974 ◯ 2086

30 5043 ◯ 4958

Arranging Numbers in Order

Exercise 13

Arrange the numbers in each row in order of size, beginning with the smallest:

1   21   37   19   29   33

_____

2   30   11   17   22   28

_____

3   37   42   39   28   45

_____

4   78   134   97   118   126

_____

5   105   86   112   97   88

_____

6   215   208   193   168   203

_____

7   881   1076   984   1048   997

_____

8 2147   3006   1984   3790   2846

_____

9 5120   4719   3857   5204   4962

_____

10 6127   5839   6083   5976   5789

_____

Exercise 14

Arrange the numbers in each row in order of size, beginning with the largest:

1   17   23   11   19   20

_____

2   36   41   29   37   46

_____

3   28   21   19   27   24

_____

4   76   101   84   97   79

_____

5   87   110   93   105   95

_____

6   126   137   154   143   139

_____

7   257   305   387   296   281

_____

8 763   1047   894   798   1053

_____

9 1143   1215   1079   1193   1258

_____

10 2742   3105   2963   2543   2851

_____

Expanded Notation

Exercise 15

Complete the following:

1 487 = ____ hundreds + ____ tens + ____ ones

2 291 = ____ hundreds + ____ tens + ____ one

3 536 = ____ hundreds + ____ tens + ____ ones

4 104 = ____ hundred + ____ tens + ____ ones

5 503 = ____ hundreds + ____ tens + ____ ones

6 670 = ____ hundreds + ____ tens + ____ ones

7 290 = ____ hundreds + ____ tens + ____ ones

8 2475 = ____ thousands + ____ hundreds + ____ tens + ____ ones

9 1832 = ____ thousand + ____ hundreds + ____ tens + ____ ones

10 3974 = ____ thousands + ____ hundreds + ____ tens + ____ ones

11 6453 = ____ thousands + ____ hundreds + ____ tens + ____ ones

12 8079 = ____ thousands + ____ hundreds + ____ tens + ____ ones

13 5042 = ____ thousands + ____ hundreds + ____ tens + ____ ones

14 3758 = ____ thousands + ____ hundreds + ____ tens + ____ ones

15 6147 = ____ thousands + ____ hundred + ____ tens + ____ ones

16 7530 = ____ thousands + ____ hundreds + ____ tens + ____ ones

17 4601 = ____ thousands + ____ hundreds + ____ tens + ____ one

18 9200 = ____ thousands + ____ hundreds + ____ tens + ____ ones

19 8005 = ____ thousands + ____ hundreds + ____ tens + ____ ones

20 9000 = ____ thousands + ____ hundreds + ____ tens + ____ ones

Exercise 16

Write the answer for each expanded form:

1   200 + 70 + 6   = _____

2   400 + 50 + 0   = _____

3   900 + 10 + 8   = _____

4   500 +   0 + 3   = _____

5   800 +   0 + 5   = _____

6 2000 + 800 + 40 + 1   = _____

7 7000 + 400 + 70 + 3   = _____

8 3000 + 900 + 50 + 2   = _____

9 6000 +   0 + 20 + 7   = _____

10 1000 + 600 + 0 + 5   = _____

11 5000 +   0 + 0 + 6   = _____

12 2000 + 500   = _____

13 3000 + 100   = _____

14 8000 +   0 + 90 + 4   = _____

15 7000 + 500 + 0 + 5   = _____

16 1000 +   0 + 0 + 6   = _____

17 5000 + 274   = _____

18 6000 + 800 + 11   = _____

19 1000 + 300 + 0 + 5   = _____

20 2000 + 500 + 80   = _____

Exercise 17

Use one of these words: thousand, thousands, hundred, hundreds, ten, tens, one, ones, to complete the following:

1   48 = 4 _____ + 8 _____

2   26 = 2 _____ + 6 _____

3   51 = 5 _____ + 1 _____

4   60 = 6 _____ + 0 _____

5   80 = 8 _____ + 0 _____

6   175 = 1 _____ + 7 _____ + 5 _____

7   356 = 3 _____ + 5 _____ + 6 _____

8   760 = 7 _____ + 6 _____ + 0 _____

9   940 = 9 _____ + 4 _____ + 0 _____

10 401 = 4 _____ + 0 _____ + 1 _____

11 203 = 2 _____ + 0 _____ + 3 _____

12 1358 = 1 _____ + 3 _____ + 5 _____ + 8 _____

13 2746 = 2 _____ + 7 _____ + 4 _____ + 6 _____

14 3071 = 3 _____ + 0 _____ + 7 _____ + 1 _____

15 8206 = 8 _____ + 2 _____ + 0 _____ + 6 _____

16 5470 = 5 _____ + 4 _____ + 7 _____ + 0 _____

17 7600 = 7 _____ + 6 _____ + 0 _____ + 0 _____

18 6000 = 6 _____ + 0 _____ + 0 _____ + 0 _____

19 4080 = 4 _____ + 0 _____ + 8 _____ + 0 _____

20 6363 = 6 _____ + 3 _____ + 6 _____ + 3 _____

Revision Exercise

1 Write in figures:

   a seven hundred and eighty   _____

   b three thousand, four hundred and sixteen _____

   c one thousand, nine hundred   _____

   d six thousand and three   _____

   e five thousand and twenty   _____

2 Write in words:

   a 604 _____

   b 1437 _____

   c 5048 _____

   d 6002 _____

   e 8300 _____

3 Divide the following numbers into two groups, odd numbers and even numbers:

        8   14   15   36   43   28   10   61   72

   Odd numbers _____

   Even numbers _____

4 In each number, what is the place value of the 6?

   a 4<u>6</u>0 _____

   b 312<u>6</u> _____

   c <u>6</u>543 _____

   d 2<u>6</u>97 _____

   e 31<u>6</u>8 _____

5 Write < or > in the circles to make each statement true:

   a 12 ◯ 8   **b** 7 ◯ 16

   c 19 ◯ 21   **d** 36 ◯ 43

   e 91 ◯ 89

6 Arrange the numbers in each row in order of size, beginning with the smallest:

a 16   27   11   9   20 _____

b 13   19   8   12   15 _____

c 18   14   22   26   17 _____

d 37   29   41   50   33 _____

e 61   58   47   39   49 _____

7 Write the answer for each expanded form:

a   500 + 60 + 2   = _____

b 3000 + 400 + 50 + 7 = _____

c 1000 + 200 + 60   = _____

d 4000 +   0 + 30 + 8 = _____

e 7000 +   0 + 0 + 9 = _____

8 Use one of these words: thousand, thousands, hundred, hundreds, ten, tens, one, ones, to complete the following:

a   76 = 7 _____ + 6 _____

b 284 = 2 _____ + 8 _____ + 4 _____

c 3157 = 3 _____ + 1 _____ + 5 _____ + 7 _____

d 2608 = 2 _____ + 6 _____ + 0 _____ + 8 _____

e 1089 = 1 _____ + 0 _____ + 8 _____ + 9 _____

9 Complete the following:

a   325 = _____ hundreds + _____ tens + _____ ones

b   870 = _____ hundreds + _____ tens + _____ ones

c 3526 = ___ thousands + ___ hundreds + ___ tens + ___ ones

d 1208 = ___ thousand + ___ hundreds + ___ tens + ___ ones

e 6400 = ___ thousands + ___ hundreds + ___ tens + ___ ones

2 Operations and Relations

Addition

Exercise 1

Work out:

1 3 + 6 = _____   **8** 5 + 5 = _____   **15** 8 + 7 = _____

2 5 + 1 = _____   **9** 3 + 5 = _____   **16** 10 + 4 = _____

3 7 + 0 = _____   **10** 6 + 4 = _____   **17** 10 + 6 = _____

4 2 + 7 = _____   **11** 9 + 8 = _____   **18** 10 + 2 = _____

5 8 + 2 = _____   **12** 7 + 6 = _____   **19** 7 + 9 = _____

6 1 + 7 = _____   **13** 5 + 7 = _____   **20** 5 + 8 = _____

7 0 + 6 = _____   **14** 9 + 9 = _____

21
$$\begin{array}{r} 8 \\ 2 \\ + 6 \\ \hline \\ \hline \end{array}$$

23
$$\begin{array}{r} 9 \\ 1 \\ + 7 \\ \hline \\ \hline \end{array}$$

25
$$\begin{array}{r} 9 \\ 5 \\ + 5 \\ \hline \\ \hline \end{array}$$

27
$$\begin{array}{r} 7 \\ 8 \\ + 1 \\ \hline \\ \hline \end{array}$$

22
$$\begin{array}{r} 4 \\ 6 \\ + 8 \\ \hline \\ \hline \end{array}$$

24
$$\begin{array}{r} 5 \\ 7 \\ + 3 \\ \hline \\ \hline \end{array}$$

26
$$\begin{array}{r} 8 \\ 5 \\ + 4 \\ \hline \\ \hline \end{array}$$

28
$$\begin{array}{r} 7 \\ 2 \\ + 3 \\ \hline \\ \hline \end{array}$$

29	6	**30**	9	**31**	6	**32**	9
	6		9		7		6
	+ 5		+ 0		+ 8		+ 5
	___		___		___		___

Exercise 2

Work out:

1	32	**6**	135	**11**	251	**16**	3146
	15		312		243		+ 2403
	+ 22		+ 421		+ 203		

2	24	**7**	412	**12**	325	**17**	5062
	15		253		461		+ 2723
	+ 40		+ 132		+ 103		

3	51	**8**	203	**13**	526	**18**	4623
	23		143		240		+ 3145
	+ 5		+ 322		+ 123		

4	20	**9**	205	**14**	415	**19**	3512
	35		142		243		+ 6280
	+ 34		+ 420		+ 201		

5	8	**10**	524	**15**	274	**20**	1257
	21		232		301		+ 6731
	+ 50		+ 113		+ 224		

Exercise 3

Work out:

1   34
    + 26
——

2   53
    + 28
——

3   28
    + 35
——

4   51
    + 39
——

5   47
    + 35
——

6   27
    + 48
——

7   39
    + 45
——

8   58
    + 27
——

9   65
    + 28
——

10   37
    + 39
——

11   26
     34
    + 13
——

12   31
     19
    + 35
——

13   43
     27
    + 18
——

14   50
     17
    + 29
——

15   26
     35
    + 17
——

16   18
     17
    + 26
——

17   38
     16
    + 19
——

18   29
     15
    + 26
——

19   34
     27
    + 18
——

20   39
     16
    + 24
——

21   27
     17
    + 29
——

22   34
     19
    + 37
——

23   28
     27
    + 35
——

24   24
     39
    + 27
——

Exercise 4

Work out:

1 248 + 570	**8** 4576 + 2309	**15** 2473 + 4078	**22** 5605 + 2798
2 465 + 471	**9** 6427 + 2546	**16** 1247 + 3854	**23** 1487 + 3988
3 164 + 492	**10** 1479 + 6318	**17** 2596 + 3978	**24** 2574 + 2827
4 357 + 590	**11** 1475 + 2238	**18** 3568 + 1874	**25** 4697 + 3509
5 236 + 691	**12** 3096 + 2649	**19** 3136 + 4986	**26** 5076 + 1948
6 2043 + 1638	**13** 4387 + 3298	**20** 2749 + 1665	**27** 6987 + 2265
7 1958 + 3026	**14** 3165 + 2497	**21** 2464 + 3579	**28** 2899 + 3708

Exercise 5

Work out:

1	4206 142 + 231	6	2013 47 216 + 312	11	3140 236 428 + 7	16	2643 128 + 431
2	2315 42 + 120	7	1245 2106 24 + 112	12	1426 254 73 + 125	17	4075 845 + 219
3	5341 4 + 122	8	5317 142 28 + 402	13	3407 265 124 + 71	18	5628 472 + 46
4	2461 204 + 112	9	6411 3 242 + 39	14	335 1443 124 + 56	19	293 1466 + 709
5	346 4122 + 120	10	7230 129 27 + 312	15	2563 207 30 + 8	20	486 278 + 6523

Addition Sequences

Example

Write the missing numbers: 18, __*21*__, 24, 27, __*30*__, __*33*__, 36

Exercise 6

Write the missing numbers in the spaces:

1 2, 4, 6, _____, _____, 12, _____

2 1, 3, 5, _____, _____, 11, _____

3 15, 20, 25, _____, _____, _____, 45

4 22, 24, 26, _____, _____, _____, 34

5 1, 4, 7, 10, _____, _____, _____

6 11, 21, _____, 41, 51, _____, _____

7 8, 16, _____, 32, 40, _____, _____

8 31, 33, 35, _____, _____, 41, _____

9 _____, 8, 12, 16, _____, _____, 28

10 _____, 6, 11, 16, 21, _____, _____

11 10, 20, 30, _____, _____, _____, 70

12 60, 63, 66, _____, _____, _____, 78

13 20, 40, 60, _____, _____, 120, _____

14 _____, 21, 32, 43, _____, _____, 76

15 _____, 25, 35, 45, _____, _____, 75

Cross-number Puzzle for Addition

Work these out and write the answers in the cross-number puzzle:

Across

A 132 + 246

C 32 + 27

E 29 + 25

F 247 + 133

G 48 + 28

H 280 + 520

I 50 + 36

J 192 + 193

L 37 + 46

N 13 + 14

O 184 + 236

Down

A 120 + 237

B 325 + 421

C 337 + 243

D 275 + 625

F 190 + 196

I 46 + 39

J 17 + 15

K 48 + 39

L 24 + 58

M 18 + 12

Problems on Addition

Exercise 7

1 Find the sum of 864, 79 and 48. _____

2 A newspaper vendor sold 436 newspapers on Friday and 580 newspapers on Saturday. How many newspapers did he sell altogether for the two days? _____

3 Mr Walkes the farmer has 96 cows, 235 sheep and 178 pigs. How many animals does he have altogether? _____

4 In a cricket match, the West Indies made 290 runs in the first innings and 315 runs in the second innings. How many runs did they make in that match? _____

5 On a bus there were 36 children, 18 men and 27 women. How many people were on the bus altogether? _____

6 A bakery sold 895 buns on Monday and 750 on Tuesday. How many buns did it sell altogether for the two days? _____

7 Yesterday, Susan planted 150 cabbage plants, 325 lettuce plants and 185 sweet-pepper plants. How many plants did she plant in all? _____

8 A farmer collected 1075 eggs in one week and 895 eggs in another week. How many eggs did he collect in the two weeks? _____

9 Last week, my mother picked 468 golden apples, 195 plums and 387 limes. How many apples, plums and limes did she pick in all? _____

10 At the seaside, Mary collected 72 shells, Susan collected 57, Jane collected 68 and Michelle collected 43. How many shells did the four girls collect in all? _____

Subtraction

Exercise 8

Complete:

1 $7 - 2$ = _____

2 $9 - 6$ = _____

3 $6 - 4$ = _____

4 $5 - 5$ = _____

5 $8 - 2$ = _____

6 $4 - 0$ = _____

7 $7 - 4$ = _____

8 $9 - 2$ = _____

9 $8 - 6$ = _____

10 $6 - 5$ = _____

11 $10 - 3$ = _____

12 $10 - 4$ = _____

13 $10 - 8$ = _____

14 $10 - 6$ = _____

15 10 – 10 = _____   **23** 14 – 8 = _____

16 12 – 5 = _____   **24** 11 – 9 = _____

17 15 – 8 = _____   **25** 12 – 7 = _____

18 13 – 7 = _____   **26** 13 – 5 = _____

19 16 – 9 = _____   **27** 16 – 9 = _____

20 18 – 9 = _____   **28** 17 – 8 = _____

21 17 – 9 = _____   **29** 14 – 5 = _____

22 15 – 7 = _____   **30** 11 – 3 = _____

 Exercise 9

Work out:

1   64
    – 21
    _____
    _____

2   87
    – 23
    _____
    _____

3   98
    – 36
    _____
    _____

4   59
    – 41
    _____
    _____

5   85
    – 30
    _____
    _____

6   76
    – 53
    _____
    _____

7   84
    – 34
    _____
    _____

8   67
    – 45
    _____
    _____

9   63
    – 10
    _____
    _____

10   98
    – 36
    _____
    _____

11   287
    – 144
    _____
    _____

12   258
    – 143
    _____
    _____

13   266
    – 123
    _____
    _____

14   286
    – 135
    _____
    _____

15   287
    – 172
    _____
    _____

16   278
    – 160
    _____
    _____

| 17 | 394
− 201 | 18 | 385
− 162 | 19 | 496
− 251 | 20 | 579
− 126 |

Exercise 10

Work out:

1	24 − 19	7	47 − 28	13	74 − 59	19	281 − 158
2	25 − 17	8	36 − 28	14	82 − 65	20	562 − 237
3	31 − 27	9	46 − 38	15	92 − 63	21	417 − 292
4	33 − 28	10	54 − 36	16	264 − 128	22	309 − 176
5	36 − 27	11	61 − 44	17	260 − 119	23	526 − 196
6	45 − 19	12	73 − 37	18	453 − 127	24	818 − 483

Exercise 11

Work out:

1	2389 − 3	**6**	4265 − 134	**11**	8567 −3416	**16**	6342 −1241
2	1578 − 6	**7**	2734 − 201	**12**	3945 −1632	**17**	7387 −2056
3	2489 − 31	**8**	6589 − 552	**13**	7869 −4315	**18**	3849 −2310
4	3567 − 60	**9**	2674 − 170	**14**	6326 −1302	**19**	5748 −3606
5	1248 − 37	**10**	6483 − 213	**15**	5476 −1324	**20**	9435 −3132

Exercise 12

Work out:

1	4123 − 7	**3**	1476 − 28	**5**	5120 − 67	**7**	5465 −2728
2	2160 − 3	**4**	2043 − 39	**6**	3742 −1258	**8**	6734 −3870

9 7186
  − 2938
_____

10 5943
  − 1396
_____

11 2864
   − 87
_____

12 3521
   − 66
_____

13 2730
   − 75
_____

14 4623
  − 149
_____

15 2704
  − 236
_____

16 3142
  − 483
_____

17 2201
  − 834
_____

18 1957
  − 968
_____

19 3400
  − 524
_____

20 4126
  − 539
_____

21 5420
  − 1723
_____

22 3916
  − 2938
_____

23 4346
  − 2547
_____

24 2460
  − 1973
_____

25 5030
  − 4937
_____

Subtraction Sequences

─Example─

Write the missing numbers:

a 20, 19, 18, 17, **16**, **15**, **14**

b 46, 45, 43, 40, **36**, **31**, **25**

Exercise 13

Write the missing numbers in each sequence:

1 45, 43, 41, _____, _____, _____, 33

2 61, 56, 51, _____, _____, _____, 31

3 40, 38, 36, 34, _____, _____, _____

4 _____, 70, 60, 50, 40, _____, _____

5 _____, 87, 76, 65, 54, _____, _____

6 81, 80, 78, 75, _____, _____, _____

7 _____, 82, 77, 72, 67, _____, _____

8 36, 35, _____, 33, 32, _____, _____

9 50, 46, _____, 38, 34, _____, _____

10 64, 54, 44, 34, _____, _____, _____

Cross-number Puzzle for Subtraction

Work these out and write the answers in the cross-number puzzle:

Across

A 48 − 15

C 472 − 121

F 375 − 153

H 82 − 18

I 213 − 133

J 492 − 309

L 96 − 52

N 100 − 42

O 823 − 219

P 480 − 345

Q 73 − 16

Down

A 49 − 17

B 92 − 60

D 86 − 30

E 72 − 58

G 507 − 224

J 256 − 105

K 995 − 112

L 633 − 228

M 630 − 183

Problems on Subtraction

Exercise 14

1 Take 62 from 200.

2 From 536 take 270.

3 What is the difference between 78 and 144?

4 What is 182 minus 96?

5 What number must be added to 47 to make 150?

6 There are 424 pupils at our school and 185 are boys. How many girls are at our school?

7

A box contains 225 bananas. 165 of them are ripe. How many bananas are not ripe?

8 What number is 17 less than 300?

9 758 athletes ran the half-marathon race and 1125 ran the ten-kilometre race. How many more athletes ran the ten-kilometre race than the half-marathon race?

10

A fish vendor bought 750 flying fish. She sold 385 of them. How many did she have left?

Multiplication

Revision

Exercise 15

Complete:

1 $4 \times 3 =$ _____

2 $2 \times 5 =$ _____

3 $8 \times 2 =$ _____

4 $7 \times 3 =$ _____

5 $4 \times 4 =$ _____

6 $5 \times 4 =$ _____

7 $8 \times 3 =$ _____

8 $9 \times 2 =$ _____

9 $6 \times 5 =$ _____

10 $3 \times 4 =$ _____

11 $0 \times 4 =$ _____

12 $9 \times 3 =$ _____

13 $5 \times 3 =$ _____

14 $7 \times 4 =$ _____

15 $8 \times 5 =$ _____

16 $7 \times 2 =$ _____

17 $9 \times 4 =$ _____

18 $5 \times 2 =$ _____

19 $3 \times 5 =$ _____

20 $6 \times 4 =$ _____

Exercise 16

Work out:

1
$$\begin{array}{r} 43 \\ \times\ 2 \\ \hline \end{array}$$

2
$$\begin{array}{r} 33 \\ \times\ 3 \\ \hline \end{array}$$

3
$$\begin{array}{r} 52 \\ \times\ 4 \\ \hline \end{array}$$

4
$$\begin{array}{r} 61 \\ \times\ 5 \\ \hline \end{array}$$

5
$$\begin{array}{r} 82 \\ \times\ 4 \\ \hline \end{array}$$

6
$$\begin{array}{r} 16 \\ \times\ 4 \\ \hline \end{array}$$

7
$$\begin{array}{r} 23 \\ \times\ 5 \\ \hline \end{array}$$

8
$$\begin{array}{r} 65 \\ \times\ 3 \\ \hline \end{array}$$

9
$$\begin{array}{r} 78 \\ \times\ 2 \\ \hline \end{array}$$

10
$$\begin{array}{r} 93 \\ \times\ 4 \\ \hline \end{array}$$

11
$$\begin{array}{r} 243 \\ \times\ \ \ 2 \\ \hline \end{array}$$

12
$$\begin{array}{r} 215 \\ \times\ \ \ 3 \\ \hline \end{array}$$

13	124 × 4	**16**	162 × 3	**19**	172 × 4	**22**	395 × 2
14	116 × 5	**17**	172 × 2	**20**	292 × 3	**23**	276 × 3
15	208 × 4	**18**	180 × 5	**21**	265 × 2	**24**	138 × 4

Multiplication by 6

Exercise 17

Complete:

1   Multiply by 6.

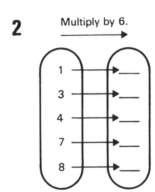

0 → 0
2
5
6
9

30
36
0
12
54

2   Multiply by 6.

1 → ___
3 → ___
4 → ___
7 → ___
8 → ___

Exercise 18

Work out:

1	41 × 6	**2**	60 × 6	**3**	81 × 6	**4**	90 × 6

5 $\begin{array}{r}70\\ \times\ 6\\ \hline\end{array}$	**11** $\begin{array}{r}32\\ \times\ 6\\ \hline\end{array}$	**17** $\begin{array}{r}87\\ \times\ 6\\ \hline\end{array}$	**23** $\begin{array}{r}305\\ \times\ 6\\ \hline\end{array}$
6 $\begin{array}{r}310\\ \times\ 6\\ \hline\end{array}$	**12** $\begin{array}{r}28\\ \times\ 6\\ \hline\end{array}$	**18** $\begin{array}{r}69\\ \times\ 6\\ \hline\end{array}$	**24** $\begin{array}{r}603\\ \times\ 6\\ \hline\end{array}$
7 $\begin{array}{r}601\\ \times\ 6\\ \hline\end{array}$	**13** $\begin{array}{r}54\\ \times\ 6\\ \hline\end{array}$	**19** $\begin{array}{r}42\\ \times\ 6\\ \hline\end{array}$	**25** $\begin{array}{r}312\\ \times\ 6\\ \hline\end{array}$
8 $\begin{array}{r}811\\ \times\ 6\\ \hline\end{array}$	**14** $\begin{array}{r}26\\ \times\ 6\\ \hline\end{array}$	**20** $\begin{array}{r}98\\ \times\ 6\\ \hline\end{array}$	**26** $\begin{array}{r}215\\ \times\ 6\\ \hline\end{array}$
9 $\begin{array}{r}16\\ \times\ 6\\ \hline\end{array}$	**15** $\begin{array}{r}73\\ \times\ 6\\ \hline\end{array}$	**21** $\begin{array}{r}407\\ \times\ 6\\ \hline\end{array}$	**27** $\begin{array}{r}240\\ \times\ 6\\ \hline\end{array}$
10 $\begin{array}{r}15\\ \times\ 6\\ \hline\end{array}$	**16** $\begin{array}{r}95\\ \times\ 6\\ \hline\end{array}$	**22** $\begin{array}{r}809\\ \times\ 6\\ \hline\end{array}$	**28** $\begin{array}{r}335\\ \times\ 6\\ \hline\end{array}$

Multiplication by 7

Exercise 19

Complete:

1 $3 \times 7 =$ _____

2 $7 \times 7 =$ _____

3 $1 \times 7 =$ _____

4 $6 \times 7 =$ _____

5 $4 \times 7 =$ _____

6 $0 \times 7 =$ _____

7 $9 \times 7 =$ _____

8 $2 \times 7 =$ _____

9 $5 \times 7 =$ _____

10 $8 \times 7 =$ _____

Exercise 20

Work out:

1
```
  41
×  7
─────
```

2
```
  60
×  7
─────
```

3
```
  81
×  7
─────
```

4
```
  50
×  7
─────
```

5
```
  90
×  7
─────
```

6
```
 110
×   7
─────
```

7
```
 101
×   7
─────
```

8
```
 711
×   7
─────
```

9
```
 206
×   7
─────
```

10
```
 608
×   7
─────
```

11
```
  12
×  7
─────
```

12
```
  13
×  7
─────
```

13
```
  73
×  7
─────
```

14
```
  64
×  7
─────
```

15
```
  56
×  7
─────
```

16
```
  35
×  7
─────
```

17
```
  66
×  7
─────
```

18
```
  74
×  7
─────
```

19
```
  98
×  7
─────
```

20
```
  63
×  7
─────
```

21
```
 240
×   7
─────
```

22
```
 350
×   7
─────
```

23
```
 812
×   7
─────
```

24
```
 215
×   7
─────
```

25
```
 425
×   7
─────
```

26
```
 372
×   7
─────
```

27
```
 284
×   7
─────
```

28
```
 537
×   7
─────
```

Multiplication by 8

Exercise 21

Complete:

1 Multiply by 8.

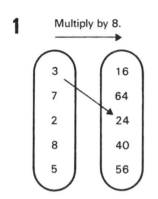

3	16
7	64
2	24
8	40
5	56

2 Multiply by 8.

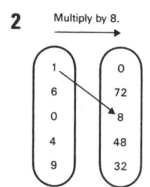

1	0
6	72
0	8
4	48
9	32

Exercise 22

Work out:

1 71
× 8
—

2 80
× 8
—

3 51
× 8
—

4 60
× 8
—

5 90
× 8
—

6 42
× 8
—

7 65
× 8
—

8 73
× 8
—

9 29
× 8
—

10 86
× 8
—

11 59
× 8
—

12 84
× 8
—

13 97
× 8
—

14 68
× 8
—

15 77
× 8
—

16 321
× 8
—

| 17 | 450
× 8 | 21 | 411
× 8 | 25 | 276
× 8 | 28 | 927
× 8 |

| 18 | 370
× 8 | 22 | 410
× 8 | 26 | 296
× 8 | 29 | 834
× 8 |

| 19 | 209
× 8 | 23 | 313
× 8 | 27 | 435
× 8 | 30 | 392
× 8 |

| 20 | 507
× 8 | 24 | 318
× 8 |

Multiplication by 9

Exercise 23

Complete:

1 $2 \times 9 =$ ____

2 $7 \times 9 =$ ____

3 $4 \times 9 =$ ____

4 $0 \times 9 =$ ____

5 $9 \times 9 =$ ____

6 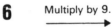 Multiply by 9.

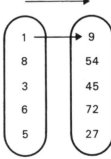

1	→	9
8		54
3		45
6		72
5		27

Exercise 24

Work out:

1	71 × 9	**9**	900 × 9	**17**	62 × 9	**24**	85 × 9
2	80 × 9	**10**	500 × 9	**18**	55 × 9	**25**	56 × 9
3	61 × 9	**11**	160 × 9	**19**	36 × 9	**26**	247 × 9
4	51 × 9	**12**	140 × 9	**20**	84 × 9	**27**	198 × 9
5	90 × 9	**13**	307 × 9	**21**	78 × 9	**28**	286 × 9
6	810 × 9	**14**	408 × 9	**22**	29 × 9	**29**	785 × 9
7	711 × 9	**15**	506 × 9	**23**	97 × 9	**30**	573 × 9
8	601 × 9	**16**	43 × 9				

Multiplication by 10

┌─────────────────────── Remember ───────────────────────┐

To multiply a number by 10, just add a zero (0) to the right-hand side of the ones figure.

$6 \times 10 = 60 \quad 28 \times 10 = 280 \quad 209 \times 10 = 2090$

└───┘

 Exercise 25

Complete:

1	$3 \times 10 =$ ____	**10**	$16 \times 10 =$ ____	**18**	$460 \times 10 =$ ____
2	$7 \times 10 =$ ____	**11**	$26 \times 10 =$ ____	**19**	$804 \times 10 =$ ____
3	$1 \times 10 =$ ____	**12**	$80 \times 10 =$ ____	**20**	$309 \times 10 =$ ____
4	$8 \times 10 =$ ____	**13**	$65 \times 10 =$ ____	**21**	$500 \times 10 =$ ____
5	$5 \times 10 =$ ____	**14**	$48 \times 10 =$ ____	**22**	$225 \times 10 =$ ____
6	$10 \times 10 =$ ____	**15**	$53 \times 10 =$ ____	**23**	$946 \times 10 =$ ____
7	$31 \times 10 =$ ____	**16**	$76 \times 10 =$ ____	**24**	$453 \times 10 =$ ____
8	$15 \times 10 =$ ____	**17**	$270 \times 10 =$ ____	**25**	$824 \times 10 =$ ____
9	$11 \times 10 =$ ____				

Multiplication by 20

┌─────────────────────── Example ───────────────────────┐

$$14 \times 20 \qquad \text{or} \qquad \begin{array}{r} 14 \\ \times\ 20 \\ \hline 280 \\ \hline \end{array}$$

$= 14 \times 10 \times 2$

$= 140 \times 2$

$= 280$

First add a zero (0), then multiply by 2.

└───┘

Exercise 26

Complete:

1 15 × 20 = _____   **6** 23 × 20 = _____   **11** 38 × 20 = _____

2 21 × 20 = _____   **7** 41 × 20 = _____   **12** 31 × 20 = _____

3 19 × 20 = _____   **8** 36 × 20 = _____   **13** 40 × 20 = _____

4 22 × 20 = _____   **9** 30 × 20 = _____   **14** 46 × 20 = _____

5 18 × 20 = _____   **10** 10 × 20 = _____   **15** 49 × 20 = _____

Multiplication by 30

─Example─

$$23 \times 30 \qquad \text{or} \qquad \begin{array}{r} 23 \\ \times\, 30 \\ \hline 690 \\ \hline \end{array}$$

$$= 23 \times 10 \times 3$$
$$= 230 \times 3$$
$$= 690$$

First add a zero (0), then multiply by 3.

Exercise 27

Complete:

1 12 × 30 = _____   **6** 27 × 30 = _____   **11** 13 × 30 = _____

2 14 × 30 = _____   **7** 29 × 30 = _____   **12** 7 × 30 = _____

3 19 × 30 = _____   **8** 31 × 30 = _____   **13** 28 × 30 = _____

4 21 × 30 = _____   **9** 8 × 30 = _____   **14** 33 × 30 = _____

5 24 × 30 = _____   **10** 10 × 30 = _____   **15** 32 × 30 = _____

Multiplication by 40

┌─ Example ─┐

$$16 \times 40 \qquad \text{or} \qquad \begin{array}{r} 16 \\ \times\, 40 \\ \hline 640 \end{array}$$
$$= 16 \times 10 \times 4$$
$$= 160 \times 4$$
$$= 640$$

First add a zero (0), then multiply by 4.

Exercise 28

Complete:

1 $7 \times 40 =$ _____   **6** $15 \times 40 =$ _____   **11** $5 \times 40 =$ _____

2 $9 \times 40 =$ _____   **7** $17 \times 40 =$ _____   **12** $21 \times 40 =$ _____

3 $10 \times 40 =$ _____   **8** $19 \times 40 =$ _____   **13** $3 \times 40 =$ _____

4 $12 \times 40 =$ _____   **9** $11 \times 40 =$ _____   **14** $22 \times 40 =$ _____

5 $8 \times 40 =$ _____   **10** $13 \times 40 =$ _____   **15** $23 \times 40 =$ _____

Multiplication by 50

┌─ Example ─┐

$$12 \times 50 \qquad \text{or} \qquad \begin{array}{r} 12 \\ \times\, 50 \\ \hline 600 \end{array}$$
$$= 12 \times 10 \times 5$$
$$= 120 \times 5$$
$$= 600$$

First add a zero (0), then multiply by 5.

Exercise 29

Complete:

1 8 × 50 = ____ **6** 4 × 50 = ____ **11** 10 × 50 = ____

2 6 × 50 = ____ **7** 7 × 50 = ____ **12** 9 × 50 = ____

3 3 × 50 = ____ **8** 16 × 50 = ____ **13** 11 × 50 = ____

4 5 × 50 = ____ **9** 14 × 50 = ____ **14** 15 × 50 = ____

5 2 × 50 = ____ **10** 13 × 50 = ____ **15** 17 × 50 = ____

Multiplication by 20, 30, 40 and 50

Exercise 30

Complete:

1 9 × 20 = ____ **11** 17 × 30 = ____ **21** 25 × 30 = ____

2 8 × 30 = ____ **12** 19 × 20 = ____ **22** 35 × 20 = ____

3 5 × 50 = ____ **13** 14 × 50 = ____ **23** 23 × 30 = ____

4 7 × 40 = ____ **14** 45 × 20 = ____ **24** 19 × 40 = ____

5 24 × 20 = ____ **15** 16 × 50 = ____ **25** 68 × 50 = ____

6 30 × 30 = ____ **16** 20 × 40 = ____ **26** 76 × 40 = ____

7 26 × 30 = ____ **17** 19 × 50 = ____ **27** 49 × 40 = ____

8 32 × 20 = ____ **18** 16 × 40 = ____ **28** 95 × 30 = ____

9 17 × 50 = ____ **19** 18 × 40 = ____ **29** 87 × 30 = ____

10 38 × 20 = ____ **20** 29 × 20 = ____ **30** 84 × 50 = ____

Multiplication by 60, 70, 80 and 90

---Example---

$$32 \times 60 \qquad \text{or} \qquad \begin{array}{r} 32 \\ \times 60 \\ \hline 1920 \end{array}$$

$$= 32 \times 10 \times 6$$
$$= 320 \times 6$$
$$= 1920$$

First add a zero (0), then multiply by 6.

Exercise 31

Work out:

1	$\begin{array}{r} 6 \\ \times 60 \\ \hline \end{array}$	**7**	$\begin{array}{r} 8 \\ \times 90 \\ \hline \end{array}$	**13**	$\begin{array}{r} 61 \\ \times 70 \\ \hline \end{array}$	**19**	$\begin{array}{r} 36 \\ \times 60 \\ \hline \end{array}$
2	$\begin{array}{r} 9 \\ \times 60 \\ \hline \end{array}$	**8**	$\begin{array}{r} 7 \\ \times 80 \\ \hline \end{array}$	**14**	$\begin{array}{r} 46 \\ \times 60 \\ \hline \end{array}$	**20**	$\begin{array}{r} 58 \\ \times 70 \\ \hline \end{array}$
3	$\begin{array}{r} 8 \\ \times 70 \\ \hline \end{array}$	**9**	$\begin{array}{r} 20 \\ \times 80 \\ \hline \end{array}$	**15**	$\begin{array}{r} 69 \\ \times 90 \\ \hline \end{array}$	**21**	$\begin{array}{r} 89 \\ \times 70 \\ \hline \end{array}$
4	$\begin{array}{r} 5 \\ \times 70 \\ \hline \end{array}$	**10**	$\begin{array}{r} 30 \\ \times 90 \\ \hline \end{array}$	**16**	$\begin{array}{r} 87 \\ \times 80 \\ \hline \end{array}$	**22**	$\begin{array}{r} 37 \\ \times 60 \\ \hline \end{array}$
5	$\begin{array}{r} 6 \\ \times 80 \\ \hline \end{array}$	**11**	$\begin{array}{r} 50 \\ \times 70 \\ \hline \end{array}$	**17**	$\begin{array}{r} 46 \\ \times 80 \\ \hline \end{array}$	**23**	$\begin{array}{r} 92 \\ \times 90 \\ \hline \end{array}$
6	$\begin{array}{r} 4 \\ \times 90 \\ \hline \end{array}$	**12**	$\begin{array}{r} 80 \\ \times 60 \\ \hline \end{array}$	**18**	$\begin{array}{r} 67 \\ \times 90 \\ \hline \end{array}$	**24**	$\begin{array}{r} 83 \\ \times 90 \\ \hline \end{array}$

25   58
      × 80
    ‾‾‾‾‾‾

    ‾‾‾‾‾‾

27   64
      × 90
    ‾‾‾‾‾‾

    ‾‾‾‾‾‾

29   59
      × 60
    ‾‾‾‾‾‾

    ‾‾‾‾‾‾

26   24
      × 90
    ‾‾‾‾‾‾

    ‾‾‾‾‾‾

28   73
      × 70
    ‾‾‾‾‾‾

    ‾‾‾‾‾‾

30   38
      × 80
    ‾‾‾‾‾‾

    ‾‾‾‾‾‾

Cross-number Puzzle for Multiplication

Work these out and write the answers in the cross-number puzzle:

Across

A 8×4

C 46×3

F 51×6

H 183×4

J 91×5

K 86×3

L 38×2

N 16×6

P 137×4

S 43×8

U 34×7

W 172×4

X 95×5

Down

A 167×2

B 41×5

D 93×4

E 167×5

G 219×3

I 41×7

M 13×5

N 117×8

O 72×9

Q 53×8

R 93×9

T 16×3

V 17×5

Problems on Multiplication

Exercise 32

1 Multiply 67 by 3. _____

2 Find the product of 80 and 9. _____

3 What is 67 times 8? _____

4

Every day John sells 65 newspapers. How many newspapers does he sell in a week (7 days)? _____

5 On every trip, a minibus carries 28 passengers. How many passengers does it carry on 6 trips? _____

6 When 5 boys shared a bag of sweets, each boy received 15 sweets. How many sweets were shared? _____

7 What number divided by 4 gives 18? _____

8 A box holds 75 sticks of chalk. How many sticks will 8 similar boxes hold? _____

9 The school canteen sells 105 packets of Corncurls every day. How many packets of Corncurls does the canteen sell in 5 days? _____

10 A textbook contains 126 pages. How many pages are there altogether in 7 copies of the same book? _____

Division

Exercise 33

Work out:

1 3)60

5 6)66

9 8)808

13 2)268

2 5)50

6 2)284

10 7)70

14 2)402

3 4)84

7 3)639

11 4)480

15 3)633

4 3)69

8 5)500

12 6)606

Exercise 34

Work out:

1 4)92

5 7)84

9 6)96

13 4)248

2 3)78

6 5)75

10 5)80

14 6)726

3 6)78

7 5)95

11 3)756

15 8)960

4 8)96

8 4)60

12 4)568

16 5)150

17 2)526   **19** 2)386   **21** 6)732   **23** 2)968

_____   _____   _____   _____

18 9)189   **20** 9)180   **22** 5)620   **24** 9)207

_____   _____   _____   _____

Exercise 35

Complete:

1 552 ÷ 2 = _____   **10** 916 ÷ 2 = _____   **18** 830 ÷ 5 = _____

2 432 ÷ 3 = _____   **11** 716 ÷ 4 = _____   **19** 616 ÷ 4 = _____

3 528 ÷ 3 = _____   **12** 822 ÷ 3 = _____   **20** 952 ÷ 7 = _____

4 738 ÷ 6 = _____   **13** 852 ÷ 6 = _____   **21** 771 ÷ 3 = _____

5 675 ÷ 5 = _____   **14** 973 ÷ 7 = _____   **22** 912 ÷ 6 = _____

6 732 ÷ 4 = _____   **15** 734 ÷ 2 = _____   **23** 376 ÷ 2 = _____

7 984 ÷ 8 = _____   **16** 912 ÷ 8 = _____   **24** 890 ÷ 5 = _____

8 875 ÷ 7 = _____   **17** 792 ÷ 6 = _____   **25** 548 ÷ 4 = _____

9 725 ÷ 5 = _____

Exercise 36

Complete:

1 29 ÷ 2 = _____   **4** 86 ÷ 4 = _____   **7** 58 ÷ 5 = _____

2 65 ÷ 3 = _____   **5** 94 ÷ 9 = _____   **8** 79 ÷ 7 = _____

3 69 ÷ 6 = _____   **6** 83 ÷ 8 = _____   **9** 81 ÷ 2 = _____

10 34 ÷ 3 = _____

11 49 ÷ 4 = _____

12 65 ÷ 2 = _____

13 84 ÷ 8 = _____

14 58 ÷ 5 = _____

15 61 ÷ 6 = _____

16 73 ÷ 4 = _____

17 66 ÷ 5 = _____

18 97 ÷ 4 = _____

19 68 ÷ 5 = _____

20 99 ÷ 7 = _____

21 938 ÷ 3 = _____

22 709 ÷ 7 = _____

23 605 ÷ 6 = _____

24 395 ÷ 3 = _____

25 809 ÷ 8 = _____

26 506 ÷ 5 = _____

27 663 ÷ 6 = _____

28 558 ÷ 5 = _____

29 485 ÷ 4 = _____

30 635 ÷ 3 = _____

31 907 ÷ 3 = _____

32 904 ÷ 9 = _____

33 847 ÷ 4 = _____

34 771 ÷ 7 = _____

35 887 ÷ 8 = _____

Exercise 37

Complete:

1 127 ÷ 6 = _____

2 158 ÷ 5 = _____

3 164 ÷ 8 = _____

4 147 ÷ 2 = _____

5 128 ÷ 3 = _____

6 154 ÷ 3 = _____

7 358 ÷ 7 = _____

8 167 ÷ 4 = _____

9 183 ÷ 9 = _____

10 107 ÷ 5 = _____

11 209 ÷ 9 = _____

12 186 ÷ 5 = _____

13 193 ÷ 4 = _____

14 289 ÷ 9 = _____

15 280 ÷ 9 = _____

16 439 ÷ 4 = _____

17 528 ÷ 5 = _____

18 217 ÷ 2 = _____

19 942 ÷ 9 = _____

20 518 ÷ 4 = _____

21 731 ÷ 2 = _____

22 943 ÷ 6 = _____

23 906 ÷ 7 = _____

24 913 ÷ 8 = _____

25 673 ÷ 5 = _____

Division as the Inverse of Multiplication

┌─────────────────── Example ───────────────────┐

$$4 \times 5 = 20$$

so $20 \div 4 =$ _5_

and $20 \div 5 =$ _4_

└──┘

 Exercise 38

Complete:

1 $8 \times 2 = 16$
so $16 \div 2 =$ _____
and $16 \div 8 =$ _____

2 $7 \times 4 = 28$
so $28 \div 4 =$ _____
and $28 \div 7 =$ _____

3 $9 \times 5 = 45$
so $45 \div 9 =$ _____
and $45 \div 5 =$ _____

4 $6 \times 7 = 42$
so $42 \div 6 =$ _____
and $42 \div 7 =$ _____

5 $8 \times 7 = 56$
so $56 \div 8 =$ _____
and $56 \div 7 =$ _____

6 $8 \times 9 = 72$
so $72 \div 9 =$ _____
and $72 \div 8 =$ _____

7 $9 \times 6 = 54$
so $54 \div 9 =$ _____
and $54 \div 6 =$ _____

8 $5 \times 6 = 30$
so $30 \div 6 =$ _____
and $30 \div 5 =$ _____

9 $4 \times 9 = 36$
so $36 \div 4 =$ _____
and $36 \div 9 =$ _____

10 $8 \times 5 = 40$
so $40 \div 8 =$ _____
and $40 \div 5 =$ _____

11 $5 \times 10 = 50$
so $50 \div 5 =$ _____
and $50 \div 10 =$ _____

12 $10 \times 7 = 70$
so $70 \div 10 =$ _____
and $70 \div 7 =$ _____

Exercise 39

Complete:

1 $9 \times 3 =$ _____
so   _____ $\div 3 = 9$
and   _____ $\div 9 = 3$

2 $6 \times 8 =$ _____
so   _____ $\div 6 = 8$
and   _____ $\div 8 = 6$

3 $7 \times 5 =$ _____
so   _____ $\div 5 = 7$
and   _____ $\div 7 = 5$

4 $7 \times 9 =$ _____
so   _____ $\div 7 = 9$
and   _____ $\div 9 = 7$

5 $4 \times 8 =$ _____
so   _____ $\div 8 = 4$
and   _____ $\div 4 = 8$

6 $4 \times 7 =$ _____
so   _____ $\div 4 = 7$
and   _____ $\div 7 = 4$

7 $8 \times 8 =$ _____
so   _____ $\div 8 = 8$

8 $6 \times 4 =$ _____
so   _____ $\div 4 = 6$
and   _____ $\div 6 = 4$

9 $8 \times 3 =$ _____
so   _____ $\div 3 = 8$
and   _____ $\div 8 = 3$

10 $7 \times 3 =$ _____
so   _____ $\div 3 = 7$
and   _____ $\div 7 = 3$

Exercise 40

Complete:

1 $8 \times$ _____ $= 16$
so $16 \div 8$ $=$ _____
and $16 \div$ _____ $= 8$

2 $3 \times$ _____ $= 12$
so $12 \div 3$ $=$ _____
and $12 \div$ _____ $= 3$

3 $2 \times$ _____ $= 18$
so $18 \div 2$ $=$ _____
and $18 \div$ _____ $= 2$

4 $10 \times$ _____ $= 50$
so $50 \div 10$ $=$ _____
and $50 \div$ _____ $= 10$

5 _____ × 9   = 27
so   27 ÷ 9   = _____
and   27 ÷ _____ = 9

8 _____ × 8   = 24
so   24 ÷ 8   = _____
and   24 ÷ _____ = 8

6 _____ × 6   = 60
so   60 ÷ 6   = _____
and   60 ÷ _____ = 6

9   8 × _____ = 80
so   80 ÷ 8   = _____
and   80 ÷ _____ = 8

7 _____ × 5   = 45
so   45 ÷ 5   = _____
and   45 ÷ _____ = 5

10 _____ × 6   = 42
so   42 ÷ 6   = _____
and   42 ÷ _____ = 6

Cross-number Puzzle for Division

Work these out and write the answers in the cross-number puzzle:

Across

A 24 ÷ 2

C 408 ÷ 3

F 756 ÷ 3

G 294 ÷ 7

H 216 ÷ 6

I 168 ÷ 8

J 102 ÷ 3

L 70 ÷ 5

N 428 ÷ 2

P 840 ÷ 3

Q 392 ÷ 7

Down

A 492 ÷ 4

B 512 ÷ 2

D 682 ÷ 2

E 310 ÷ 5

I 968 ÷ 4

K 876 ÷ 6

L 24 ÷ 2

M 96 ÷ 2

O 60 ÷ 4

Problems on Division

Exercise 41

1 Divide 156 by 6. _____

2 How many fours are there in 120? _____

3 How many nines are there in 261? _____

4 How many pupils can each receive 5 plums from a bag
containing 225 plums? _____

5 Share 144 marbles equally among 4 boys. How many
does each boy receive? _____

6 2 boys share 56 markers equally. How many will each
receive? _____

7 There are 100 sweet biscuits in a box. How many
children can get 4 sweet biscuits each? _____

8

A book contains 130 pages. If I read the same number
of pages each night for 5 nights and finish the book,
how many pages do I read each night? _____

9 161 chairs are placed in 7 equal rows. How many chairs are
there in each row? _____

10 How many packets each with 6 pencils can be filled
from a box of 144 pencils? _____

Special Properties

The Commutative Property for Addition and Multiplication

> **─Example─**
> **a** *The commutative property for addition*: $4 + 6 = 6 + 4$
> **b** *The commutative property for multiplication*: $3 \times 5 = 5 \times 3$

Exercise 42

Complete:

1 $3 + 5 = 5 +$ ____

2 $1 + 7 = 7 +$ ____

3 $6 + 3 =$ ____ $+ 6$

4 $2 + 4 =$ ____ $+ 2$

5 ____ $+ 8 = 8 + 3$

6 $9 +$ ____ $= 4 + 9$

7 $6 +$ ____ $= 7 + 6$

8 $9 + 8 = 8 +$ ____

9 ____ $+ 3 = 3 + 9$

10 $7 +$ ____ $= 8 + 7$

11 $2 \times 6 = 6 \times$ ____

12 $3 \times 4 = 4 \times$ ____

13 $7 \times 2 =$ ____ $\times 7$

14 $2 \times 4 = 4 \times$ ____

15 $5 \times 4 =$ ____ $\times 5$

16 $8 \times 2 = 2 \times$ ____

17 ____ $\times 3 = 3 \times 7$

18 ____ $\times 8 = 8 \times 4$

19 ____ $\times 6 = 6 \times 5$

20 $2 \times$ ____ $= 8 \times 2$

The Associative Property for Addition and Multiplication

> **─Example a─**
> **i** $(8 + 9) + 2$ or **ii** $(8 + 2) + 9$
> $= 17 + 2$ $\qquad\qquad = 10 + 9$
> $= 19$ $\qquad\qquad\quad = 19$
> So $(8 + 9) + 2 = (8 + 2) + 9$
> Combining as in part ii makes addition easier.

┌─────────────────── **Example b** ───────────────────┐
│   i $(76 \times 5) \times 2$ or   **ii** $76 \times (5 \times 2)$ │
│   = 380×2 = 76×10 │
│   = 760   = 760 │
│   So $(76 \times 5) \times 2 = 76 \times (5 \times 2)$ │
│   Combining as in part ii makes multiplication easier. │
└──┘

Exercise 43

Combine the following to make addition or multiplication easier, then work out the answers:

1 $7 + 8 + 3 =$ _____ = _____

2 $9 + 5 + 5 =$ _____ = _____

3 $9 + 7 + 1 =$ _____ = _____

4 $5 + 8 + 2 =$ _____ = _____

5 $9 + 9 + 1 =$ _____ = _____

6 $8 + 4 + 6 =$ _____ = _____

7 $19 + 7 + 1 =$ _____ = _____

8 $17 + 6 + 3 =$ _____ = _____

9 $18 + 5 + 2 =$ _____ = _____

10 $2 + 28 + 6 =$ _____ = _____

11 $8 + 39 + 1 =$ _____ = _____

12 $28 + 30 + 2 =$ _____ = _____

13 $76 + 4 + 9 =$ _____ = _____

14 $60 + 17 + 40 =$ _____ = _____

15 $50 + 8 + 50 =$ _____ = _____

16 $5 \times 13 \times 2 =$ _____ = _____

17 $16 \times 5 \times 2 =$ _____ = _____

18 $5 \times 27 \times 2 =$ _____ = _____

19 $2 \times 9 \times 50 =$ _____ = _____

20 $4 \times 8 \times 25 =$ _____ = _____

21 $4 \times 25 \times 9 =$ _____ = _____

22 $5 \times 25 \times 4 =$ _____ = _____

23 $2 \times 19 \times 2 =$ _____ = _____

24 $3 \times 26 \times 2 =$ _____ = _____

25 $3 \times 16 \times 2 =$ _____ = _____

26 $18 \times 6 \times 5 =$ _____ = _____

27 $46 \times 8 \times 5 =$ _____ = _____

28 $15 \times 7 \times 2 =$ _____ = _____

29 $12 \times 15 \times 4 =$ _____ = _____

30 $8 \times 17 \times 5 =$ _____ = _____

Special Properties of Zero and One

Identity property of *zero under addition*:
$$7 + 0 = 7 \qquad 0 + 12 = 12$$

Identity property of *one under multiplication*:
$$9 \times 1 = 9 \qquad 74 \times 1 = 74 \qquad 605 \times 1 = 605$$

The *product of zero* and any other number is zero:
$$15 \times 0 = 0 \qquad 27 \times 0 = 0 \qquad 0 \times 9 = 0$$

If *zero is subtracted* from any number the answer is that number:
$$7 - 0 = 7 \qquad 14 - 0 = 14 \qquad 38 - 0 = 38$$

If a number is *subtracted from the same number* the answer is zero:
$$5 - 5 = 0 \qquad 16 - 16 = 0 \qquad 64 - 64 = 0$$

Exercise 44

Complete:

1	4×0 = _____	**11**	1×18 = _____	**21**	15×1 = _____
2	7×1 = _____	**12**	$36 - 0$ = _____	**22**	24×0 = _____
3	$9 - 0$ = _____	**13**	$19 + 0$ = _____	**23**	$18 + 0$ = _____
4	$13 - 13$ = _____	**14**	$17 - 17$ = _____	**24**	$63 - 0$ = _____
5	0×6 = _____	**15**	28×0 = _____	**25**	0×11 = _____
6	$5 + 0$ = _____	**16**	0×72 = _____	**26**	$43 + 0$ = _____
7	$11 - 0$ = _____	**17**	31×1 = _____	**27**	$38 - 38$ = _____
8	$9 - 9$ = _____	**18**	1×26 = _____	**28**	1×47 = _____
9	12×0 = _____	**19**	$42 - 42$ = _____	**29**	$0 + 34$ = _____
10	$14 + 0$ = _____	**20**	$26 - 0$ = _____	**30**	$49 - 49$ = _____

Revision Exercise

1 Work these out:

a	46 + 23	**b**	235 + 740	**c**	37 + 59	**d**	86 + 77
e	456 + 138	**f**	394 + 483	**g**	275 + 326	**h**	213 46 480 + 22
i	4378 + 1034	**j**	3827 + 4519				

2 Work these out:

a   564
  − 210
	b 73
	  − 29
		c 7642
		 − 5016
			d 341
			  − 126

e 608 − 273	**f** 4632 − 8	**g** 3145 − 27	**h** 8230 − 168
i 5416 − 1836	**j** 6213 − 2567		

3 a What number is 10 less than 96? _____

   b Find the difference between 84 and 200. _____

   c

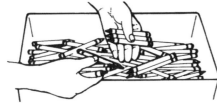

In a box there are 74 red crayons, 87 yellow crayons and 165 green crayons. How many crayons are there altogether? _____

   d At our school, there are 75 girls and 58 boys. How many more girls than boys are at our school? _____

   e Find the sum of 340, 69 and 79. _____

4 Work these out:

a 83 × 3	**b** 607 × 5	**c** 89 × 7	**d** 216 × 6
e 740 × 9	**f** 358 × 8	**g** 87 × 10	**h** 65 × 60
i 73 × 80	**j** 69 × 70		

5 Complete:

a $648 \div 2 =$ _____   b $96 \div 4 =$ _____   c $735 \div 7 =$ _____

d $126 \div 6 =$ _____   e $800 \div 5 =$ _____   f $912 \div 8 =$ _____

g $641 \div 6 =$ _____   h $857 \div 2 =$ _____   i $862 \div 3 =$ _____

j $715 \div 4 =$ _____

6 Write the missing numbers in the boxes:

a $6 \times \boxed{} = 48$ b $\boxed{} \times 5 = 35$

c $\boxed{} \div 4 = 8$ d $\boxed{} \div 8 = 9$

e $\boxed{} \div 6 = 9$

7 Complete:

a $6 \times 1 =$ _____   b $7 \times 0 =$ _____

c $10 - 0 =$ _____   d $9 - 9 =$ _____

e $12 + 0 =$ _____   f $13 - 13 =$ _____

g $4 - 0 =$ _____   h $1 \times 16 =$ _____

i $0 + 11 =$ _____   j $0 \times 18 =$ _____

8 a 84 chairs are arranged in 7 equal rows. How many chairs are in each row? _____

b Oranges are packed 6 in a box. How many boxes are needed to pack 150 oranges? _____

c At the school fair, Debbie sold 28 hot dogs. Her teacher sold 8 times as many. How many hot dogs did her teacher sell? _____

d Share 325 sea shells equally among 5 tourists. How many sea shells would each tourist receive? _____

e Find the product of 63 and 9. _____

9 Write the missing numbers:

a 4, 8, 12, _____, 20

b 37, 32, 27, 22, _____

c _____, 18, 27, 36, 45

d 6, 12, _____, 24, 30

e 11, 21, 31, 41, _____

3 Fractions

Equivalent Fractions

Halves, Quarters, Eighths and Sixteenths

The diagram shows that one-half equals two-quarters.

This is also written as

$$\frac{1}{2} = \frac{2}{4}$$

$\frac{1}{2}$	
$\frac{1}{4}$	$\frac{1}{4}$

Exercise 1

Use the diagram below to complete the following:

1 WHOLE															
$\frac{1}{2}$								$\frac{1}{2}$							
$\frac{1}{4}$				$\frac{1}{4}$				$\frac{1}{4}$				$\frac{1}{4}$			
$\frac{1}{8}$		$\frac{1}{8}$		$\frac{1}{8}$		$\frac{1}{8}$		$\frac{1}{8}$		$\frac{1}{8}$		$\frac{1}{8}$		$\frac{1}{8}$	
$\frac{1}{16}$	$\frac{1}{16}$	$\frac{1}{16}$	$\frac{1}{16}$	$\frac{1}{16}$	$\frac{1}{16}$	$\frac{1}{16}$	$\frac{1}{16}$	$\frac{1}{16}$	$\frac{1}{16}$	$\frac{1}{16}$	$\frac{1}{16}$	$\frac{1}{16}$	$\frac{1}{16}$	$\frac{1}{16}$	$\frac{1}{16}$

1 $\frac{1}{2} = \frac{}{8}$

2 $\frac{1}{2} = \frac{}{16}$

3 $\frac{3}{8} = \frac{}{16}$

4 $\frac{1}{8} = \frac{}{16}$

5 $\frac{3}{4} = \frac{}{8}$

6 $\frac{3}{4} = \frac{}{16}$

7 $\frac{5}{8} = \frac{}{16}$

8 $\frac{1}{4} = \frac{}{16}$

9 $\frac{2}{4} = \frac{}{8}$

10 $\frac{2}{4} = \frac{}{16}$

11 $\frac{3}{4} = \frac{}{16}$

12 $\frac{1}{4} = \frac{}{8}$

13 $\frac{2}{8} = \frac{}{16}$ **14** $1 = \frac{}{4}$ **15** $1 = \frac{}{8}$

Exercise 2

Use the diagrams that follow to complete these:

1 WHOLE			
$\frac{1}{4}$	$\frac{1}{4}$	$\frac{1}{4}$	$\frac{1}{4}$

1 WHOLE							
$\frac{1}{8}$	$\frac{1}{8}$	$\frac{1}{8}$	$\frac{1}{8}$	$\frac{1}{8}$	$\frac{1}{8}$	$\frac{1}{8}$	$\frac{1}{8}$

1 $\frac{2}{4} + \frac{1}{4} = $ _____

2 $\frac{1}{4} + \frac{1}{4} + \frac{1}{4} = $ _____

3 $\frac{3}{4} - \frac{1}{4} = $ _____

4 $\frac{3}{4} - \frac{2}{4} = $ _____

5 $\frac{2}{4} - \frac{1}{4} = $ _____

6 $1 - \frac{3}{4} = $ _____

7 $1 - \frac{1}{4} = $ _____

8 $\frac{1}{8} + \frac{1}{8} + \frac{1}{8} = $ _____

9 $\frac{3}{8} + \frac{1}{8} = $ _____

10 $\frac{5}{8} + \frac{1}{8} + \frac{1}{8} = $ _____

11 $\frac{5}{8} + \frac{2}{8} = $ _____

12 $\frac{1}{8} + \frac{7}{8} = $ _____

13 $\frac{3}{8} - \frac{1}{8} = $ _____

14 $\frac{5}{8} - \frac{4}{8} = $ _____

15 $\frac{7}{8} - \frac{3}{8} = $ _____

16 $1 - \frac{3}{8} = $ _____

17 $1 - \frac{7}{8} = $ _____

18 $1 - \frac{1}{8} = $ _____

19 $1 - \frac{5}{8} = $ _____

1 WHOLE															
$\frac{1}{16}$	$\frac{1}{16}$	$\frac{1}{16}$	$\frac{1}{16}$	$\frac{1}{16}$	$\frac{1}{16}$	$\frac{1}{16}$	$\frac{1}{16}$	$\frac{1}{16}$	$\frac{1}{16}$	$\frac{1}{16}$	$\frac{1}{16}$	$\frac{1}{16}$	$\frac{1}{16}$	$\frac{1}{16}$	$\frac{1}{16}$

20 $\frac{1}{16} + \frac{1}{16} + \frac{1}{16} =$ _____

21 $\frac{2}{16} + \frac{1}{16} + \frac{2}{16} =$ _____

22 $\frac{5}{16} + \frac{4}{16} =$ _____

23 $\frac{1}{16} + \frac{3}{16} + \frac{5}{16} =$ _____

24 $\frac{7}{16} + \frac{3}{16} + \frac{1}{16} =$ _____

25 $\frac{7}{16} - \frac{1}{16} =$ _____

26 $\frac{5}{16} - \frac{3}{16} =$ _____

27 $\frac{9}{16} - \frac{4}{16} =$ _____

28 $\frac{3}{16} - \frac{1}{16} =$ _____

29 $\frac{11}{16} - \frac{7}{16} =$ _____

30 $\frac{8}{16} - \frac{3}{16} =$ _____

31 $\frac{13}{16} - \frac{11}{16} =$ _____

32 $1 - \frac{7}{16} =$ _____

33 $1 - \frac{5}{16} =$ _____

34 $1 - \frac{9}{16} =$ _____

35 $1 - \frac{3}{16} =$ _____

Thirds, Sixths and Twelfths

The diagram shows that two-thirds equal four-sixths.

This is also written as

$$\frac{2}{3} = \frac{4}{6}$$

$\frac{1}{3}$		$\frac{1}{3}$	
$\frac{1}{6}$	$\frac{1}{6}$	$\frac{1}{6}$	$\frac{1}{6}$

Exercise 3

Use the diagram below to complete the following:

1 WHOLE											
$\frac{1}{3}$				$\frac{1}{3}$				$\frac{1}{3}$			
$\frac{1}{6}$		$\frac{1}{6}$		$\frac{1}{6}$		$\frac{1}{6}$		$\frac{1}{6}$		$\frac{1}{6}$	
$\frac{1}{12}$	$\frac{1}{12}$	$\frac{1}{12}$	$\frac{1}{12}$	$\frac{1}{12}$	$\frac{1}{12}$	$\frac{1}{12}$	$\frac{1}{12}$	$\frac{1}{12}$	$\frac{1}{12}$	$\frac{1}{12}$	$\frac{1}{12}$

1 $\frac{1}{3} = \frac{}{6}$　　　　**4** $\frac{1}{3} = \frac{}{12}$　　　　**7** $1 = \frac{}{6}$

2 $\frac{3}{6} = \frac{}{12}$　　　　**5** $\frac{2}{3} = \frac{}{12}$　　　　**8** $1 = \frac{}{12}$

3 $\frac{4}{6} = \frac{}{12}$　　　　**6** $1 = \frac{}{3}$　　　　**9** $\frac{5}{6} = \frac{}{12}$

Fifths and Tenths

The diagram shows that
one-fifth equals two-tenths.

This is also written as

$$\frac{1}{5} = \frac{2}{10}$$

$\frac{1}{5}$	
$\frac{1}{10}$	$\frac{1}{10}$

Exercise 4

Use the diagram below to complete the following:

1 WHOLE									
$\frac{1}{5}$		$\frac{1}{5}$		$\frac{1}{5}$		$\frac{1}{5}$		$\frac{1}{5}$	
$\frac{1}{10}$	$\frac{1}{10}$	$\frac{1}{10}$	$\frac{1}{10}$	$\frac{1}{10}$	$\frac{1}{10}$	$\frac{1}{10}$	$\frac{1}{10}$	$\frac{1}{10}$	$\frac{1}{10}$

1 $\frac{3}{5} = \frac{}{10}$　　　　**3** $\frac{2}{5} = \frac{}{10}$　　　　**5** $1 = \frac{}{10}$

2 $\frac{4}{5} = \frac{}{10}$　　　　**4** $1 = \frac{}{5}$

Addition and Subtraction of Fractions

The diagram shows

$$\frac{2}{6} + \frac{2}{6} + \frac{1}{6} = \frac{5}{6}$$

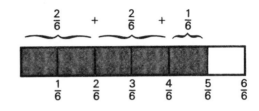

Exercise 5

Use the diagrams that follow to complete these:

1 WHOLE					
$\frac{1}{6}$	$\frac{1}{6}$	$\frac{1}{6}$	$\frac{1}{6}$	$\frac{1}{6}$	$\frac{1}{6}$

$$\frac{1}{6} \quad \frac{2}{6} \quad \frac{3}{6} \quad \frac{4}{6} \quad \frac{5}{6} \quad \frac{6}{6}$$

1 $\frac{1}{6} + \frac{1}{6} =$ _____

2 $\frac{1}{6} + \frac{1}{6} + \frac{1}{6} =$ _____

3 $\frac{2}{6} + \frac{2}{6} =$ _____

4 $\frac{3}{6} + \frac{2}{6} =$ _____

5 $\frac{1}{6} + \frac{2}{6} + \frac{2}{6} =$ _____

6 $\frac{5}{6} - \frac{1}{6} =$ _____

7 $\frac{4}{6} - \frac{3}{6} =$ _____

8 $\frac{5}{6} - \frac{3}{6} =$ _____

9 $\frac{3}{6} - \frac{1}{6} =$ _____

10 $1 - \frac{5}{6} =$ _____

11 $1 - \frac{1}{6} =$ _____

12 $1 - \frac{4}{6} =$ _____

1 WHOLE											
$\frac{1}{12}$	$\frac{1}{12}$	$\frac{1}{12}$	$\frac{1}{12}$	$\frac{1}{12}$	$\frac{1}{12}$	$\frac{1}{12}$	$\frac{1}{12}$	$\frac{1}{12}$	$\frac{1}{12}$	$\frac{1}{12}$	$\frac{1}{12}$

$\frac{1}{12}$ $\frac{2}{12}$ $\frac{3}{12}$ $\frac{4}{12}$ $\frac{5}{12}$ $\frac{6}{12}$ $\frac{7}{12}$ $\frac{8}{12}$ $\frac{9}{12}$ $\frac{10}{12}$ $\frac{11}{12}$ $\frac{12}{12}$

13 $\frac{1}{12} + \frac{1}{12} =$ _____

14 $\frac{2}{12} + \frac{1}{12} + \frac{1}{12} =$ _____

15 $\frac{1}{12} + \frac{4}{12} =$ _____

16 $\frac{3}{12} + \frac{4}{12} =$ _____

17 $\frac{5}{12} + \frac{3}{12} =$ _____

18 $\frac{9}{12} + \frac{1}{12} =$ _____

19 $\frac{3}{12} + \frac{1}{12} + \frac{5}{12} =$ _____

20 $\frac{1}{12} + \frac{5}{12} + \frac{5}{12} =$ _____

21 $\frac{1}{12} + \frac{1}{12} + \frac{3}{12} =$ _____

22 $\frac{3}{12} + \frac{2}{12} + \frac{1}{12} =$ _____

23 $\frac{4}{12} - \frac{3}{12} =$ _____

24 $\frac{7}{12} - \frac{5}{12} =$ _____

25 $\frac{5}{12} - \frac{4}{12} =$ _____

26 $\frac{9}{12} - \frac{5}{12} =$ _____

27 $\frac{11}{12} - \frac{5}{12} =$ _____

28 $\frac{6}{12} - \frac{1}{12} =$ _____

29 $1 - \frac{3}{12} =$ _____

30 $1 - \frac{5}{12} =$ _____

31 $1 - \frac{9}{12} =$ _____

32 $1 - \frac{11}{12} =$ _____

33 $\frac{3}{12} + \frac{1}{12} + \frac{3}{12} =$ _____

34 $\frac{7}{12} + \frac{5}{12} =$ _____

35 $\frac{9}{12} + \frac{2}{12} =$ _____

Exercise 6

Use the diagrams below to complete the following:

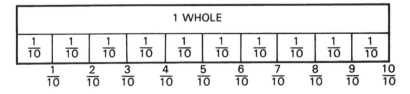

1 $\dfrac{1}{5} + \dfrac{1}{5} = $ _____

2 $\dfrac{2}{5} + \dfrac{1}{5} = $ _____

3 $\dfrac{1}{5} + \dfrac{1}{5} + \dfrac{1}{5} = $ _____

4 $\dfrac{3}{5} + \dfrac{1}{5} = $ _____

5 $\dfrac{2}{5} + \dfrac{1}{5} + \dfrac{1}{5} = $ _____

6 $\dfrac{3}{5} + \dfrac{2}{5} = $ _____

7 $\dfrac{3}{5} - \dfrac{2}{5} = $ _____

8 $\dfrac{2}{5} - \dfrac{1}{5} = $ _____

9 $\dfrac{4}{5} - \dfrac{3}{5} = $ _____

10 $\dfrac{4}{5} - \dfrac{2}{5} = $ _____

11 $1 - \dfrac{3}{5} = $ _____

12 $\dfrac{1}{10} + \dfrac{1}{10} = $ _____

13 $\dfrac{1}{10} + \dfrac{1}{10} + \dfrac{1}{10} = $ _____

14 $\dfrac{3}{10} + \dfrac{2}{10} = $ _____

15 $\dfrac{5}{10} + \dfrac{1}{10} + \dfrac{1}{10} = $ _____

16 $\dfrac{7}{10} + \dfrac{2}{10} = $ _____

17 $\dfrac{3}{10} - \dfrac{2}{10} = $ _____

18 $\dfrac{5}{10} - \dfrac{1}{10} = $ _____

19 $\dfrac{6}{10} - \dfrac{2}{10} = $ _____

20 $1 - \dfrac{5}{10} = $ _____

21 $1 - \dfrac{7}{10} = $ _____

22 $1 - \dfrac{4}{10} = $ _____

Comparison of Fractions

1 WHOLE							
$\frac{1}{2}$				$\frac{1}{2}$			
$\frac{1}{4}$		$\frac{1}{4}$		$\frac{1}{4}$		$\frac{1}{4}$	
$\frac{1}{8}$	$\frac{1}{8}$	$\frac{1}{8}$	$\frac{1}{8}$	$\frac{1}{8}$	$\frac{1}{8}$	$\frac{1}{8}$	$\frac{1}{8}$
$\frac{1}{16}$ $\frac{1}{16}$	$\frac{1}{16}$ $\frac{1}{16}$	$\frac{1}{16}$ $\frac{1}{16}$	$\frac{1}{16}$ $\frac{1}{16}$	$\frac{1}{16}$ $\frac{1}{16}$	$\frac{1}{16}$ $\frac{1}{16}$	$\frac{1}{16}$ $\frac{1}{16}$	$\frac{1}{16}$ $\frac{1}{16}$

From the diagram, it can be seen that $\frac{1}{2}$ is greater than $\frac{1}{4}$, or $\frac{1}{4}$ is less than $\frac{1}{2}$. This can be written as

$$\frac{1}{2} \ > \ \frac{1}{4} \ \text{or} \ \frac{1}{4} \ < \ \frac{1}{2}$$

Exercise 7

With the help of the diagram above and the diagrams that follow, write < (is less than),
> (is greater than), or = (is equal to) in each circle to make each statement true:

1 $\frac{1}{4}$ ◯ $\frac{1}{2}$ **6** $\frac{1}{2}$ ◯ $\frac{5}{16}$ **11** $\frac{3}{16}$ ◯ $\frac{1}{4}$

2 $\frac{2}{8}$ ◯ $\frac{1}{8}$ **7** $\frac{3}{4}$ ◯ $\frac{1}{2}$ **12** $\frac{5}{8}$ ◯ $\frac{10}{16}$

3 $\frac{1}{2}$ ◯ $\frac{3}{8}$ **8** $\frac{5}{16}$ ◯ $\frac{3}{4}$ **13** $\frac{1}{16}$ ◯ $\frac{1}{8}$

4 $\frac{5}{8}$ ◯ $\frac{1}{2}$ **9** $\frac{4}{8}$ ◯ $\frac{1}{2}$ **14** $\frac{7}{8}$ ◯ $\frac{7}{16}$

5 $\frac{1}{4}$ ◯ $\frac{3}{8}$ **10** $\frac{8}{16}$ ◯ $\frac{3}{4}$ **15** $\frac{2}{4}$ ◯ $\frac{4}{8}$

1 WHOLE											
$\frac{1}{3}$				$\frac{1}{3}$				$\frac{1}{3}$			
$\frac{1}{6}$		$\frac{1}{6}$		$\frac{1}{6}$		$\frac{1}{6}$		$\frac{1}{6}$		$\frac{1}{6}$	
$\frac{1}{12}$	$\frac{1}{12}$	$\frac{1}{12}$	$\frac{1}{12}$	$\frac{1}{12}$	$\frac{1}{12}$	$\frac{1}{12}$	$\frac{1}{12}$	$\frac{1}{12}$	$\frac{1}{12}$	$\frac{1}{12}$	$\frac{1}{12}$

16 $\frac{1}{6}$ $\bigcirc$ $\frac{1}{3}$ **21** $\frac{2}{6}$ $\bigcirc$ $\frac{1}{3}$

17 $\frac{2}{3}$ $\bigcirc$ $\frac{2}{6}$ **22** $\frac{3}{12}$ $\bigcirc$ $\frac{1}{6}$

18 $\frac{5}{6}$ $\bigcirc$ $\frac{7}{12}$ **23** $\frac{6}{12}$ $\bigcirc$ $\frac{5}{6}$

19 $\frac{1}{3}$ $\bigcirc$ $\frac{4}{12}$ **24** $\frac{2}{3}$ $\bigcirc$ $\frac{8}{12}$

20 $\frac{9}{12}$ $\bigcirc$ $\frac{2}{3}$ **25** $\frac{1}{6}$ $\bigcirc$ $\frac{1}{12}$

1 WHOLE									
$\frac{1}{5}$		$\frac{1}{5}$		$\frac{1}{5}$		$\frac{1}{5}$		$\frac{1}{5}$	
$\frac{1}{10}$	$\frac{1}{10}$	$\frac{1}{10}$	$\frac{1}{10}$	$\frac{1}{10}$	$\frac{1}{10}$	$\frac{1}{10}$	$\frac{1}{10}$	$\frac{1}{10}$	$\frac{1}{10}$

26 $\frac{1}{5}$ $\bigcirc$ $\frac{1}{10}$ **29** $\frac{4}{5}$ $\bigcirc$ $\frac{7}{10}$

27 $\frac{2}{5}$ $\bigcirc$ $\frac{3}{10}$ **30** $\frac{3}{5}$ $\bigcirc$ $\frac{5}{10}$

28 $\frac{6}{10}$ $\bigcirc$ $\frac{3}{5}$

Fractions of Sets and Shapes

Finding a Fraction of a Set of Objects

Exercise 8

Complete the following for the set of coins:

1 $\frac{1}{2}$ of them = _____

2 $\frac{1}{4}$ of them = _____

3 $\frac{1}{8}$ of them = _____

4 $\frac{1}{16}$ of them = _____

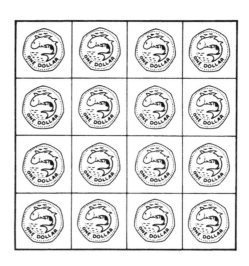

Complete the following for the set of leaves:

5 $\frac{1}{2}$ of them = _____

6 $\frac{1}{3}$ of them = _____

7 $\frac{1}{6}$ of them = _____

8 $\frac{1}{12}$ of them = _____

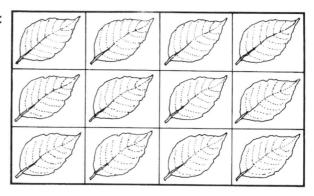

Complete the following for the set of apples:

9 $\frac{1}{5}$ of them = _____

10 $\frac{1}{10}$ of them = _____

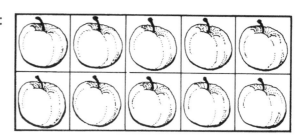

Shading a Fraction of a Set or Shape

Exercise 9

1 Colour $\frac{1}{2}$ of this set:

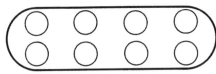

2 Colour $\frac{1}{4}$ of this set:

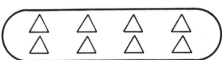

3 Colour $\frac{1}{3}$ of this set:

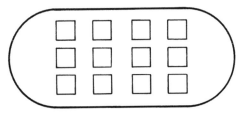

4 Colour $\frac{1}{6}$ of this set:

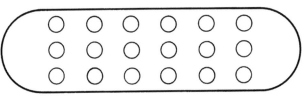

5 Colour $\frac{1}{5}$ of this set:

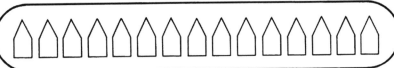

6 Colour $\frac{3}{8}$:

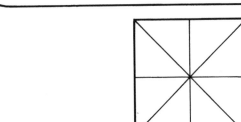

7 Colour $\frac{7}{10}$:

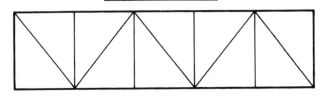

8 Colour $\frac{2}{3}$:

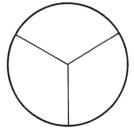

9 Colour $\frac{3}{4}$:

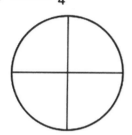

10 Colour $\frac{5}{12}$:

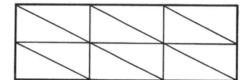

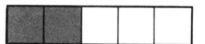

Recognising Fractions of a Shape or Set

Exercise 10

For each shape, or set, write the fraction that is   **a** shaded,   **b** not shaded:

1 a Shaded _____
   b Not shaded _____

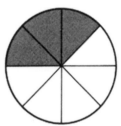

2 a Shaded _____
   b Not shaded _____

3 a Shaded _____
   b Not shaded _____

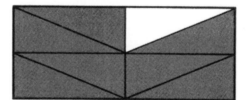

4 a Shaded _____
   b Not shaded _____

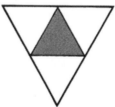

5 a Shaded _____
   b Not shaded _____

6 a Shaded _____
  b Not shaded _____

7 a Shaded _____
  b Not shaded _____

8 a Shaded _____
  b Not shaded _____

9 a Shaded _____
  b Not shaded _____

10 a Shaded _____
   b Not shaded _____

Finding Fractions of Quantities

Exercise 11

Complete:

1 $\frac{1}{2}$ of 12 = _____

2 $\frac{1}{2}$ of 18 = _____

3 $\frac{1}{2}$ of 24 = _____

4 $\frac{1}{2}$ of 60 minutes = _____

5 $\frac{1}{2}$ of 100 grams = _____

6 $\frac{1}{3}$ of 6 = _____

7 $\frac{1}{3}$ of 15 = _____

8 $\frac{1}{3}$ of 27 = _____

9 $\frac{1}{3}$ of 12 eggs = _____

10 $\frac{1}{3}$ of 75 cents = _____

11 $\frac{1}{4}$ of 8 = _____

12 $\frac{1}{4}$ of 20 = _____

13 $\frac{1}{4}$ of 32 = _____

14 $\frac{1}{4}$ of 60 seconds = _____

15 $\frac{1}{4}$ of 100 cents = _____

16 $\frac{1}{5}$ of 10 = _____

17 $\frac{1}{5}$ of 15 = _____

18 $\frac{1}{5}$ of 30 = _____

19 $\frac{1}{5}$ of 45 marbles = _____

20 $\frac{1}{5}$ of 60 centimetres = _____

21 $\frac{1}{6}$ of 12 = _____

22 $\frac{1}{6}$ of 24 = _____

23 $\frac{1}{6}$ of 30 = _____

24 $\frac{1}{6}$ of 60 apples = _____

25 $\frac{1}{6}$ of 90 litres = _____

26 $\frac{1}{8}$ of 8 = _____

27 $\frac{1}{8}$ of 16 = _____

28 $\frac{1}{8}$ of 40 = _____

29 $\frac{1}{8}$ of 72 metres = _____

30 $\frac{1}{8}$ of 80 biscuits = _____

31 $\frac{1}{10}$ of 10 = _____

32 $\frac{1}{10}$ of 30 = _____

33 $\frac{1}{10}$ of 40 = _____

34 $\frac{1}{10}$ of 70 kilograms = _____

35 $\frac{1}{10}$ of 100 millilitres = _____

Problem Solving

Exercise 12

1 Mother divided her birthday cake into 8 equal pieces. She gave 5 pieces to friends.

   a What fraction did she give to friends? _____

   b What fraction did she have left? _____

2 There are 10 members in our science club, 3 girls and 7 boys:

   a What fraction of the members are boys? _____

   b What fraction of the members are girls? _____

3 $\frac{3}{5}$ of the pupils in our class like mathematics. What fraction do not like mathematics? _____

4 Aunt Con divided a bar of chocolate into 3 equal pieces. She ate 1 piece and gave 2 pieces to her niece.

   a What fraction did she eat? _____

   b What fraction did she give to her niece? _____

5 My grandfather divided a piece of land into 4 equal plots. He gave 1 plot to his daughter and the rest to his son.

   a What fraction did he give to his daughter? _____

   b What fraction did he give to his son? _____

6 Our sports club has 8 cricket balls, of which 3 are red and 5 are white:

   a What fraction of the balls are white? _____

   b What fraction of the balls are red? _____

7 In a group of 10 girls, 7 like netball and 3 like volleyball:

   a What fraction of the girls like netball? _____

   b What fraction of the girls like volleyball? _____

8 André had 10 dollars. He spent 7 dollars on toys.

   a What fraction of his money did he spend on toys?   _____

   b What fraction of his money did he have left?   _____

9 Ann had 6 golden apples. She ate 5 of them.

   a What fraction did she eat?   _____

   b What fraction did she have left?   _____

10 In a box, there were 12 eggs. 5 of them were broken.

   a What fraction were broken?   _____

   b What fraction were not broken?   _____

Revision Exercise

1 Complete the following:

  a $\dfrac{1}{2} = \dfrac{}{6}$ **b** $\dfrac{1}{5} = \dfrac{}{10}$ **c** $\dfrac{2}{4} = \dfrac{}{8}$

  d $\dfrac{7}{8} = \dfrac{}{16}$ **e** $\dfrac{2}{4} = \dfrac{}{16}$ **f** $\dfrac{12}{16} = \dfrac{}{4}$

  g $\dfrac{}{5} = \dfrac{6}{10}$ **h** $\dfrac{4}{6} = \dfrac{}{12}$ **i** $\dfrac{4}{12} = \dfrac{}{3}$

  j $\dfrac{}{10} = \dfrac{1}{5}$

2 Add the following fractions:

  a $\dfrac{2}{5} + \dfrac{2}{5} = $ _____   **b** $\dfrac{1}{8} + \dfrac{2}{8} = $ _____

  c $\dfrac{2}{8} + \dfrac{1}{8} + \dfrac{2}{8} = $ _____   **d** $\dfrac{3}{8} + \dfrac{3}{8} + \dfrac{1}{8} = $ _____

  e $\dfrac{2}{12} + \dfrac{1}{12} = $ _____   **f** $\dfrac{2}{12} + \dfrac{7}{12} = $ _____

  g $\dfrac{3}{12} + \dfrac{4}{12} = $ _____   **h** $\dfrac{2}{12} + \dfrac{2}{12} + \dfrac{1}{12} = $ _____

  i $\dfrac{5}{12} + \dfrac{4}{12} = $ _____   **j** $\dfrac{3}{12} + \dfrac{4}{12} + \dfrac{2}{12} = $ _____

  k $\dfrac{1}{16} + \dfrac{2}{16} = $ _____   **l** $\dfrac{4}{16} + \dfrac{3}{16} = $ _____

m $\dfrac{3}{16} + \dfrac{1}{16} + \dfrac{3}{16} =$ _____

n $\dfrac{5}{16} + \dfrac{4}{16} =$ _____

o $\dfrac{5}{16} + \dfrac{5}{16} + \dfrac{3}{16} =$ _____

p $\dfrac{1}{6} + \dfrac{2}{6} =$ _____

q $\dfrac{2}{6} + \dfrac{1}{6} + \dfrac{2}{6} =$ _____

r $\dfrac{3}{6} + \dfrac{3}{6} =$ _____

s $\dfrac{5}{12} + \dfrac{5}{12} + \dfrac{1}{12} =$ _____

t $\dfrac{7}{16} + \dfrac{3}{16} + \dfrac{5}{16} =$ _____

3 Subtract the following fractions:

a $\dfrac{2}{6} - \dfrac{1}{6} =$ _____

b $\dfrac{3}{6} - \dfrac{1}{6} =$ _____

c $\dfrac{5}{6} - \dfrac{2}{6} =$ _____

d $\dfrac{3}{5} - \dfrac{1}{5} =$ _____

e $\dfrac{3}{8} - \dfrac{2}{8} =$ _____

f $\dfrac{6}{8} - \dfrac{1}{8} =$ _____

g $\dfrac{7}{8} - \dfrac{5}{8} =$ _____

h $\dfrac{5}{8} - \dfrac{2}{8} =$ _____

i $\dfrac{3}{12} - \dfrac{2}{12} =$ _____

j $\dfrac{5}{12} - \dfrac{3}{12} =$ _____

k $\dfrac{9}{12} - \dfrac{4}{12} =$ _____

l $\dfrac{11}{12} - \dfrac{8}{12} =$ _____

m $\dfrac{7}{16} - \dfrac{4}{16} =$ _____

n $\dfrac{8}{16} - \dfrac{3}{16} =$ _____

o $\dfrac{13}{16} - \dfrac{6}{16} =$ _____

p $1 - \dfrac{1}{5} =$ _____

q $1 - \dfrac{1}{6} =$ _____

r $1 - \dfrac{7}{12} =$ _____

s $1 - \dfrac{3}{16} =$ _____

t $1 - \dfrac{13}{16} =$ _____

4 Complete:

a $\frac{1}{2}$ of 6 = _____

b $\frac{1}{2}$ of 10 = _____

c $\frac{1}{4}$ of 16 sweets =

_____

d $\frac{1}{3}$ of 60 minutes =

_____

e $\frac{1}{8}$ of 16 = _____

f $\frac{1}{4}$ of 12 = _____

g $\frac{1}{6}$ of 48 = _____

h $\frac{1}{10}$ of 50 = _____

i $\frac{1}{10}$ of 20 litres = _____

j $\frac{1}{2}$ of 32 marbles = _____

k $\frac{1}{3}$ of 36 cents = _____

l $\frac{1}{2}$ of 50 metres = _____

m $\frac{1}{5}$ of 40 = _____

n $\frac{1}{6}$ of 36 = _____

o $\frac{1}{4}$ of 40 = _____

p $\frac{1}{5}$ of 30 grams = _____

q $\frac{1}{12}$ of 24 litres = _____

r $\frac{1}{12}$ of 60 minutes = _____

s $\frac{1}{5}$ of 50 = _____

t $\frac{1}{5}$ of 75 = _____

5 Write <, > or = in each circle to make each statement true:

a $\frac{1}{2}$ ◯ $\frac{5}{8}$ b $\frac{3}{8}$ ◯ $\frac{1}{4}$ c $\frac{1}{3}$ ◯ $\frac{2}{6}$

d $\frac{5}{6}$ ◯ $\frac{2}{3}$ e $\frac{3}{4}$ ◯ $\frac{7}{8}$ f $\frac{1}{10}$ ◯ $\frac{1}{5}$

g $\frac{3}{10}$ ◯ $\frac{5}{10}$ h $\frac{9}{10}$ ◯ $\frac{4}{5}$ i $\frac{1}{12}$ ◯ $\frac{1}{3}$

j $\frac{1}{6}$ ◯ $\frac{2}{12}$ k $\frac{8}{12}$ ◯ $\frac{2}{3}$ l $\frac{5}{16}$ ◯ $\frac{3}{16}$

m $\frac{9}{16}$ ◯ $\frac{7}{8}$ n $\frac{1}{4}$ ◯ $\frac{4}{16}$ o $\frac{1}{2}$ ◯ $\frac{7}{16}$

p $\frac{1}{2}$ ◯ $\frac{3}{4}$ q $\frac{5}{8}$ ◯ $\frac{7}{8}$ r $\frac{2}{5}$ ◯ $\frac{1}{2}$

s $\frac{8}{16}$ ◯ $\frac{1}{2}$ t $\frac{5}{6}$ ◯ $\frac{5}{12}$

6 a In my family there are 2 adults and 3 children:

    i What fraction of my family are adults? _____

    ii What fraction of my family are children? _____

b A bag contains 12 marbles, of which 5 are blue and 7 are green:

    i What fraction of the marbles are blue? _____

    ii What fraction of the marbles are green? _____

c The diagram contains 5 shapes:

    i What fraction are triangles? _____

    ii What fraction are circles? _____

    iii What fraction are rectangles? _____

d My godmother gave me a box of 6 handkerchiefs for my birthday. 3 were yellow, 2 were red and 1 was white.

    i What fraction of the handkerchiefs were yellow? _____

    ii What fraction of the handkerchiefs were red? _____

    iii What fraction of the handkerchiefs were white? _____

e There were 8 horses in a field. 3 of them were grey.

    i What fraction of the horses were grey? _____

    ii What fraction of the horses were not grey? _____

4 Measurement and Geometry

Length

Exercise 1

Measure the following lines and record their lengths in the table:

A ——————————————————

B ——————————

C ——————————

D ——————————————

E ————

F ——————————

G ————

H ——————————————

I ——————————————

J ——

Line	Measurement
A	cm
B	cm
C	cm
D	cm
E	cm
F	cm
G	cm
H	cm
I	cm
J	cm

Writing Measurements to the Nearest 5 cm

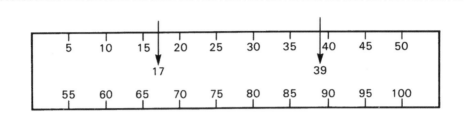

17 cm lies between 15 cm and 20 cm.
17 cm lies nearer to 15 cm than to 20 cm.
So we say that 17 cm, to the nearest 5 cm, is 15 cm.

39 cm lies between 35 cm and 40 cm.
39 cm lies nearer to 40 cm than to 35 cm.
So we say that 39 cm, to the nearest 5 cm, is 40 cm.

Exercise 2

Write the following measurements to the nearest 5 cm:

1 8 cm = _____ cm

2 21 cm = _____ cm

3 6 cm = _____ cm

4 14 cm = _____ cm

5 36 cm = _____ cm

6 23 cm = _____ cm

7 32 cm = _____ cm

8 48 cm = _____ cm

9 41 cm = _____ cm

10 29 cm = _____ cm

11 52 cm = _____ cm

12 68 cm = _____ cm

13 59 cm = _____ cm

14 63 cm = _____ cm

15 27 cm = _____ cm

16 83 cm = _____ cm

17 71 cm = _____ cm

18 94 cm = _____ cm

19 99 cm = _____ cm

20 87 cm = _____ cm

Writing Measurements to the Nearest 10 cm

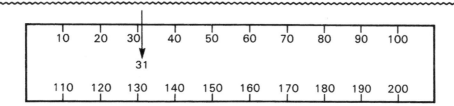

31 cm lies between 30 cm and 40 cm.
31 cm lies nearer to 30 cm than to 40 cm.
So, we say that 31 cm, to the nearest 10 cm, is 30 cm.

 Exercise 3

Write the following measurements to the nearest 10 cm:

1   18 cm = _____ cm

2   12 cm = _____ cm

3   36 cm = _____ cm

4   43 cm = _____ cm

5   59 cm = _____ cm

6   47 cm = _____ cm

7   22 cm = _____ cm

8   34 cm = _____ cm

9   71 cm = _____ cm

10   78 cm = _____ cm

11   106 cm = _____ cm

12   113 cm = _____ cm

13   147 cm = _____ cm

14   182 cm = _____ cm

15   97 cm = _____ cm

16   86 cm = _____ cm

17   92 cm = _____ cm

18   104 cm = _____ cm

19   151 cm = _____ cm

20   138 cm = _____ cm

21   122 cm = _____ cm

22   163 cm = _____ cm

23   176 cm = _____ cm

24   194 cm = _____ cm

25   198 cm = _____ cm

Comparing Measurements

Exercise 4

1 Mary is 145 cm tall. Susan is 129 cm tall. Who is taller? _____

2 Mark is 152 cm tall. Ron is 138 cm tall. Who is shorter? _____

3 Ann is 160 cm tall. Sharon is 7 cm shorter. How tall is
Sharon?   _____

Here are the heights in centimetres of a family of five:

Mr Rock   Mrs Rock   Adrian   Michelle   Crystal
170 cm   164 cm   106 cm   126 cm   85 cm

4 Who is the tallest?   _____

5 Who is the shortest?   _____

6 Who is half as tall as Mr Rock?   _____

7 Who is 38 cm shorter than Mrs Rock?   _____

8 Who is 21 cm taller than Crystal?   _____

9 How many more centimetres tall is Mr Rock than Michelle? _____

10 How many more centimetres tall is Mrs Rock than
Crystal?   _____

Capacity

Two units used for measuring capacity are the litre and the millilitre:

1000 millilitres = 1 litre

1000 mℓ = 1 ℓ

Exercise 5

1 Complete:

a _____ mℓ = $\frac{1}{2}$ ℓ

b _____ mℓ = $\frac{1}{4}$ ℓ

2

$\frac{1}{2}$ ℓ	$\frac{1}{2}$ ℓ

1 ℓ

How many half-litres of milk would it take to fill a litre container? _____

3

$\frac{1}{4}$ ℓ	$\frac{1}{4}$ ℓ	$\frac{1}{4}$ ℓ	$\frac{1}{4}$ ℓ

1 ℓ

How many quarter-litres of milk would it take to fill a litre container? _____

4 How many quarter-litres of milk would it take to fill a half-litre container? _____

Exercise 6

Here are five bottles of orange juice with their capacities in millilitres:

125 ml	250 ml	375 ml	500 ml	750 ml
A	B	C	D	E

1 Which bottle contains more, B or C? _____

2 Which bottle contains less, D or E? _____

3 How many millilitres less does C contain than D? _____

4 How many more millilitres does D contain than A? _____

5 Which bottle contains the least orange juice? _____

6 Which bottle contains the most orange juice? _____

7 Which bottle contains twice as much orange juice as A? _____

8 Which bottle contains half as much orange juice as D? _____

9 How many times can I fill A from C? _____

10 How many times can I fill A from D? _____

11 How many size B bottles would it take to fill D? _____

12 How many more millilitres of orange juice are there in C than in B? _____

13 How many size A bottles would it take to fill E? _____

Complete:

14 Bottle _____ can be filled once with A and B together.

15 Bottle _____ can be filled once with A and C together.

Mass

Two units used for weighing are the gram and the kilogram:

1000 grams = 1 kilogram

1000 g = 1 kg

Comparing Weights

Exercise 7

Here are six parcels with their weights in kilograms:

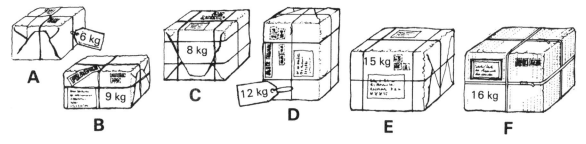

1 Which parcel is heavier, B or C? _____

2 Which parcel is heavier, F or D? _____

3 Which parcel is lighter, A or C? _____

4 Which parcel weighs less, E or D? _____

5 Which parcel is twice as heavy as C? _____

6 Which parcel is half as heavy as D? _____

7 Which parcel is 4 kg heavier than C? _____

8 Which parcel is 3 kg lighter than D? _____

9 How many more kilograms does F weigh than B? _____

Complete:

10 Parcel A is _____ kg less than parcel B.

Addition of Weights

Exercise 8

Add these:

1
```
    g
  250
+ 125
------
```

2
```
    g
  380
+ 100
------
```

3
```
    g
  125
+  70
------
```

4
```
    g
  136
+ 342
------
```

5
```
    g
  252
+ 306
------
```

6
```
   kg
  256
+ 341
------
```

7
```
   kg
  412
+ 275
------
```

8
```
   kg
  386
+ 142
------
```

9
```
   kg
  493
+ 225
------
```

10
```
   kg
  674
+ 295
------
```

11
```
   kg
  276
+ 104
------
```

12
```
   kg
  358
+ 215
------
```

13
```
   kg
  487
+ 168
------
```

14
```
   kg
  273
+ 257
------
```

15
```
   kg
  149
+ 374
------
```

Subtraction of Weights

Exercise 9

Subtract these:

1
```
    g
  563
- 241
------
```

2
```
    g
  678
- 104
------
```

3
```
    g
  854
- 234
------
```

4
```
    g
  957
- 323
------
```

	g		kg		kg		kg
5	643 − 510	**8**	426 − 294	**11**	561 − 238	**14**	810 − 264

	kg		kg		kg		kg
6	647 − 280	**9**	837 − 692	**12**	743 − 207	**15**	503 − 129

	kg		kg		kg
7	532 − 171	**10**	480 − 136	**13**	721 − 265

Multiplication of Weights

Exercise 10

Work these out:

	g		g		g		g
1	120 × 2	**4**	105 × 2	**7**	265 × 2	**10**	146 × 6

	g		g		g		kg
2	125 × 3	**5**	300 × 3	**8**	183 × 5	**11**	123 × 7

	g		g		g		kg
3	150 × 3	**6**	150 × 4	**9**	175 × 5	**12**	116 × 8

```
        kg                 kg                 kg                 kg
13     105          17    165          20    103          23    145
    ×    9              ×    6              ×    9              ×    5
   _____           _____           _____           _____


   _____           _____           _____           _____
        kg                 kg                 kg                 kg
14     107          18    136          21    143          24    164
    ×    8              ×    6              ×    6              ×    4
   _____           _____           _____           _____


   _____           _____           _____           _____
        kg                 kg                 kg                 kg
15     125          19    120          22    126          25    275
    ×    5              ×    8              ×    7              ×    3
   _____           _____           _____           _____


   _____           _____           _____           _____
        kg
16     254
    ×    3
   _____


   _____
```

Division of Weights

Exercise 11

Work these out:

1 2⟌400 g   **4** 2⟌800 g   **7** 3⟌930 g   **10** 4⟌480 g

   ____   ____   ____   ____

2 2⟌460 g   **5** 3⟌690 g   **8** 3⟌900 g   **11** 4⟌800 g

   ____   ____   ____   ____

3 2⟌640 g   **6** 3⟌360 g   **9** 4⟌840 g   **12** 4⟌880 g

   ____   ____   ____   ____

13 5) 550 g   **18** 2) 786 g   **23** 5) 835 g   **27** 6) 930 g

_____   _____   _____   _____

14 6) 600 g   **19** 3) 750 g   **24** 5) 720 g   **28** 7) 924 g

_____   _____   _____   _____

15 7) 770 g   **20** 3) 846 g   **25** 6) 732 g   **29** 8) 984 g

_____   _____   _____   _____

16 8) 880 g   **21** 4) 520 g   **26** 6) 804 g   **30** 8) 992 g

_____   _____   _____   _____

17 2) 550 g   **22** 4) 648 g

_____   _____

Problems Involving Weight

Exercise 12

1 Find the total weight of three parcels that weigh 12 kg, 10 kg and 15 kg.   _____

2 One bag of potatoes weighs 22 kg and another bag weighs 12 kg. What is the difference in weight between the two bags of potatoes?   _____

3 Susan weighs 35 kg and June weighs 28 kg. What is the sum of their weights?   _____

4 Mother bought four packets of cheese, each weighing 175 g. What is the total weight of the cheese?   _____

5 Peter cut a piece of meat weighing 900 g into four equal pieces. What was the weight of each piece?   _____

6 Kim weighs 39 kg. Kim's mother weighs 32 kg more than Kim. What is the weight of Kim's mother?   _____

7 What weight must I add to 225 g to make 420 g?   _____

8 A supermarket bought six bags of carrots. If 25 kg of carrots were in each bag, how many kilograms of carrots did the supermarket buy?   _____

9 A box of sweets weighs 750 g. How much would half the amount of sweets weigh?   _____

10 Here are the weights of four parcels:

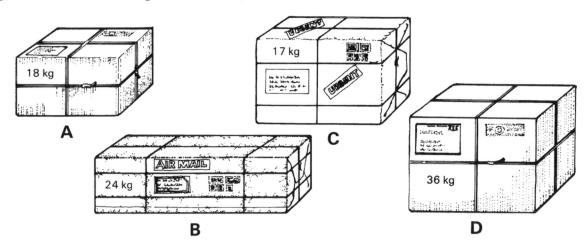

a What is the difference in weight between the smallest and largest parcels?   _____

b What is the total weight of the four parcels?   _____

c Which parcel is 7 kg less than parcel B?   _____

d Which parcel is twice as heavy as parcel A?   _____

e How many more kilograms does parcel D weigh than parcel B?   _____

Time

Exercise 13

What is the time shown by each clock?

1

_____

2

_____

3
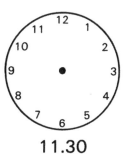
_____

4

_____

Draw hands on the clock faces to show the given times:

5
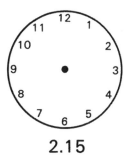
4.45

7

11.30

6
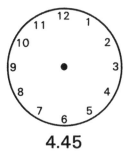
8 o'clock

8
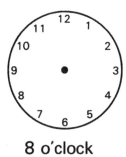
2.15

9.05　　　　9.10　　　　9.20　　　　9.25

At 5 minutes past the hour, the long hand is on 1.

At 10 minutes past the hour, the long hand is on 2.

At 20 minutes past the hour, the long hand is on 4.

At 25 minutes past the hour, the long hand is on 5.

Exercise 14

What is the time shown by each clock?

1 　**2** 　**3** 　**4**

_____　_____　_____　_____

Draw hands on the clock faces to show the given times:

5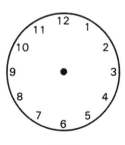

3.05
or 5 minutes past 3

6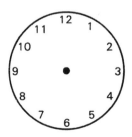

9.20
or 20 minutes past 9

7

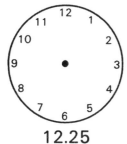

12.25
or 25 minutes past 12

8

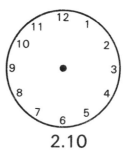

2.10
or 10 minutes past 2

3.35

At 35 minutes past the hour, the long hand is on 7.

12.40

At 40 minutes past the hour, the long hand is on 8.

7.50

At 50 minutes past the hour, the long hand is on 10.

11.55

At 55 minutes past the hour, the long hand is on 11.

Exercise 15

What is the time shown by each clock?

1

_____

2

_____

3

_____

4

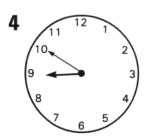

_____

Draw hands on the clock faces to show the given times:

5

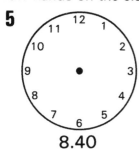

8.40

6

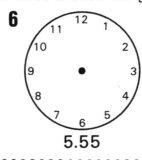

5.55

7

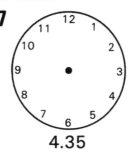

4.35

8

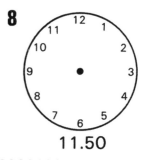

11.50

Exercise 16

What is the time shown by each clock?

1

5

9

2

6

10

3

7

11

4

8

12

Draw hands on the clock faces to show the given times:

13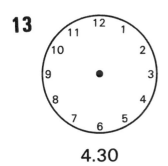

4.30

17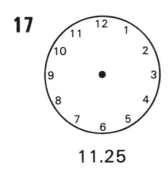

11.25

21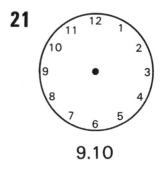

9.10

14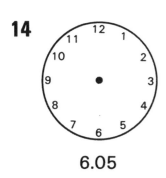

6.05

18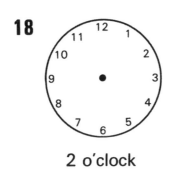

2 o'clock

22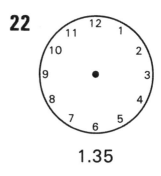

1.35

15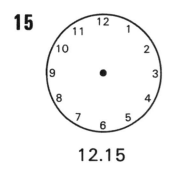

12.15

19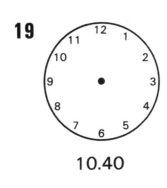

10.40

23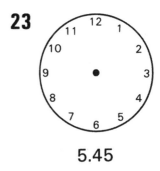

5.45

16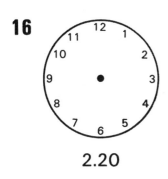

2.20

20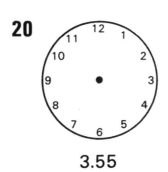

3.55

24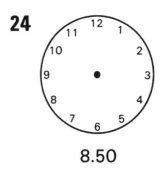

8.50

Exercise 17

Clock A shows 2 o'clock:

A

Draw in the hands and write the times for the following clocks:

1 1 hour later
than clock A

_____

2 2 hours earlier
than clock A

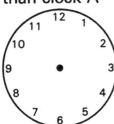

_____

3 4 hours later
than clock A

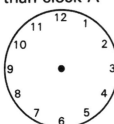

_____

4 5 hours earlier
than clock A

_____

5 30 minutes later
than clock A

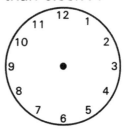

_____

6 15 minutes earlier
than clock A

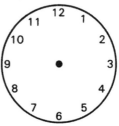

_____

7 15 minutes later
than clock A

_____

8 30 minutes earlier
than clock A

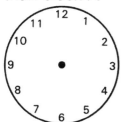

_____

9 40 minutes later
than clock A

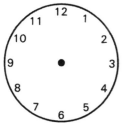

_____

10 10 minutes earlier
than clock A

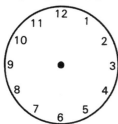

_____

11 25 minutes later
than clock A

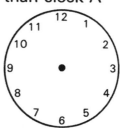

_____

12 20 minutes later
than clock A

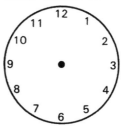

_____

13 20 minutes earlier
than clock A

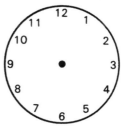

_____

14 5 minutes later
than clock A

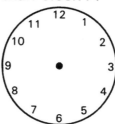

_____

Exercise 18

How much time has passed? (Each pair of clocks shows times in the same day.)

1

_____

2

_____

3

_____

4

_____

5

_____

6

from

to

_____

7

from

to

_____

8

from

to

_____

9

from

to

_____

10

from

to

_____

Exercise 19

How much time has passed? (Each pair of clocks shows times in the same day.)

1

_____

2

_____

3

_____

4

_____

5

_____

6

from

to

_____

7

from

to

_____

8

from

to

_____

9

from

to

_____

10

from

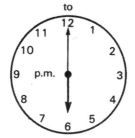
to

_____

Exercise 20

1 Colette left home at 8.00 a.m. She got to
school at 8.30 a.m. How long did she take to
get to school?

_____

2 Carlo started his homework at 7.30 p.m.
He finished at 9.00 p.m. How long did he
take to finish his homework?

_____

3 Mother started to wash the clothes at
9.30 a.m. She completed the washing at
10.30 a.m. How long did she take to wash
the clothes?

_____

4 Lunch-time for the pupils began at
12.00 a.m. and ended at 1.30 p.m.
How long did lunch-time last?

_____

5 A jet plane left Barbados at 3.30 p.m.
It arrived in New York at 8.00 p.m.
How long did the journey take?

_____

6 Our school starts at 9.00 a.m. and finishes
at 3.00 p.m. How long is school-time?

_____

7 My father started a job at 10.15 a.m. and
finished it at 12.15 a.m. How long did the
job take?

_____

8 Ann left home at 7.00 a.m. She returned
30 minutes later. At what time did she
return?

_____

9 Susan got to work at 7.30 a.m. and was
1 hour early. At what time should she get
to work?

_____

10 On Saturday, cricket begins at 1.00 p.m.
and finishes at 5.30 p.m. How long does
cricket last?

_____

Money

Exercise 21

Write the correct values in the circles:

Use: (1 ¢) (5 ¢) (10 ¢) or (25 ¢)

1 ( ) + ( ) + ( ) = (12 ¢)

2 ( ) + ( ) = (15 ¢)

3 ( ) + ( ) + ( ) = (31 ¢)

4 ( ) + ( ) = (35 ¢)

5 ( ) + ( ) + ( ) = (55 ¢)

6 ( ) + ( ) + ( ) + ( ) = (41 ¢)

7 ( ) + ( ) + ( ) + ( ) = (65 ¢)

8 ( ) + ( ) + ( ) + ( ) = (80 ¢)

9 ( ) + ( ) + ( ) + ( ) = (76 ¢)

10 ( ) + ( ) + ( ) + ( ) = (85 ¢)

Exercise 22

Write the correct values in the boxes:

Use: \$1   \$2   \$5   or   \$10

1 ☐ + ☐ + ☐ = \$4

2 ☐ + ☐ + ☐ = \$7

3 ☐ + ☐ + ☐ = \$9

4 ☐ + ☐ + ☐ = \$12

5 ☐ + ☐ + ☐ = \$16

6 ☐ + ☐ + ☐ = \$13

7 ☐ + ☐ + ☐ = \$15

8 ☐ + ☐ + ☐ = \$17

9 ☐ + ☐ + ☐ = \$20

10 ☐ + ☐ + ☐ = \$14

Exercise 23

Write the correct values in the boxes:

Use: | $10 | | $5 | | $2 | | $1 | | 50 ¢ |

| 25 ¢ | | 10 ¢ | | 5 ¢ | or | 1 ¢ |

1 ☐ + ☐ + ☐ = $2.50

2 ☐ + ☐ + ☐ + ☐ = $3.15

3 ☐ + ☐ + ☐ + ☐ = $4.30

4 ☐ + ☐ + ☐ + ☐ = $5.27

5 ☐ + ☐ + ☐ + ☐ = $7.35

6 ☐ + ☐ + ☐ + ☐ = $8.05

7 ☐ + ☐ + ☐ + ☐ = $9.10

8 ☐ + ☐ + ☐ + ☐ = $10.31

9 ☐ + ☐ + ☐ + ☐ = $10.20

10 ☐ + ☐ + ☐ + ☐ = $10.55

Exercise 24

Complete the table as shown in the example:

	$10	$5	$2	$1	25 ¢	10 ¢	5 ¢	1 ¢	Amount
Example		1	1		2	1		4	$7.64
1									$3.50
2									$4.15
3									$6.28
4									$5.43
5									$7.53
6									$8.60
7									$8.95
8									$9.30
9									$9.77
10									$10.86
11									$11.50
12									$15.74
13									$19.47
14									$21.38
15									$29.93

Exercise 25

Write the amounts as shown in the examples:

	$10	$5	$2	$1	25 ¢	10 ¢	5 ¢	1 ¢	Amount
Example	1		1	1	1	1	1		*$13.40*
Example	1	1	1		3	1		2	*$17.87*
1		1	1		1			1	
2		2	2	1		2	1		
3	1			1	2			4	
4	1	1			1	1	1		
5		1	2			2		3	
6	1		1	1	1		1	2	
7		1		2	4				
8	1		1	3	1		1	4	
9		2	2			1	2	1	
10	1	1		1	2	1		3	
11	1	1	2		1		1	2	
12	2			3		4	1		
13	2	1	2	1	1	2		1	
14	3		1	4	2		3	2	
15	3	1	2	2	1	3			

Exercise 26

Here are some items that are sold in our school shop:

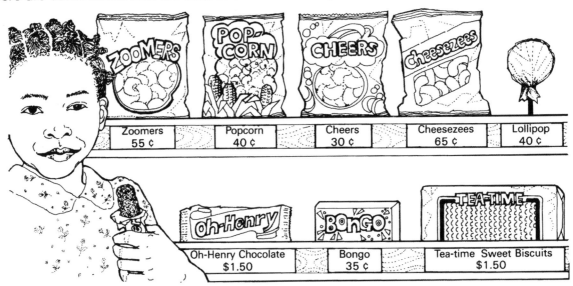

Zoomers	Popcorn	Cheers	Cheesezees	Lollipop
55 ¢	40 ¢	30 ¢	65 ¢	40 ¢

Oh-Henry Chocolate	Bongo	Tea-time Sweet Biscuits
$1.50	35 ¢	$1.50

1 Susan wants to buy an Oh-Henry Chocolate and a packet of Cheesezees. How much money would she need? _____

2 Michael bought a packet of tea-time sweet biscuits and a packet of Zoomers. How much money did he spend? _____

3 What is the total cost of a Bongo, a packet of popcorn and a packet of Zoomers? _____

4 Find the total cost of a packet of Cheesezees, a lollipop and an Oh-Henry Chocolate. _____

5 Ron bought four packets of tea-time sweet biscuits. How much money did he spend? _____

6 How much more does a packet of tea-time sweet biscuits cost than a packet of Cheesezees? _____

7 Sonia had $5.00. She bought an Oh-Henry Chocolate. How much money did she have left? _____

8 What change must I receive from $5.00, after buying three packets of Cheesezees? _____

9 What change must I receive from $5.00, after buying three lollipops and three packets of Cheers? _____

10 I had $10.00. I bought two packets of tea-time sweet
biscuits and four packets of Cheesezees.   _____

   a How much money did I spend?   _____

   b How much money did I have left?   _____

Exercise 27

Here are some items that are sold in the mini-mart:

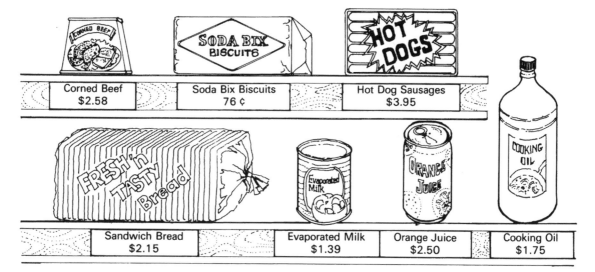

| Corned Beef $2.58 | Soda Bix Biscuits 76 ¢ | Hot Dog Sausages $3.95 |

| Sandwich Bread $2.15 | Evaporated Milk $1.39 | Orange Juice $2.50 | Cooking Oil $1.75 |

1 What is the total cost of a tin of evaporated milk and a tin of
corned beef?   _____

2 Find the total cost of a sandwich bread and a tin of hot
dog sausages.   _____

3 Mother wants to buy three tins of evaporated milk.
How much money would she need?   _____

4 How much more does a packet of hot dog sausages cost
than a tin of evaporated milk?   _____

5 Janice had $10.00. She bought a tin of corned beef.
How much money did she have left?   _____

6 I had $10.00. I bought a bottle of cooking oil and a tin
of corned beef. How much money did I have left?   _____

7 How much change is left out of $20.00 after buying
four tins of hot dog sausages?   _____

8 How much change is left out of $20.00 after buying five cans of orange juice?   _____

9 Jean bought three packets of Soda Bix Biscuits and a tin of hot dog sausages:

   a How much money did she spend?   _____

   b What change did she have left out of $10.00?   _____

10 Mother had $20.00. She bought a tin of evaporated milk, two sandwich breads and a tin of corned beef.

   a How much money did she spend?   _____

   b How much money did she have left?   _____

Shapes

Exercise 28

Write the name of each shape:

1
_____

4
_____

7
_____

2
_____

5
_____

8
_____

3
_____

6
_____

Exercise 29

Choose one of these names: cone, cube, cuboid, cylinder, sphere, to identify each of these shapes:

1

3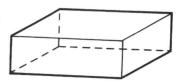

5

_____   _____   _____

2

4

_____   _____

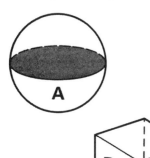

Exercise 30

Identify the shapes and complete the table:

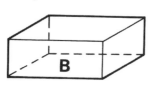
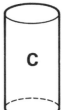
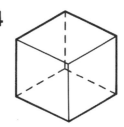

Shape	Letter
Circle	
Cone	
Cube	
Cuboid	
Cylinder	
Rectangle	
Sphere	
Square	
Triangle	

Exercise 31

1 Here is a cylinder closed at both ends:

How many faces does it have?   _____

2 How many faces does a cube have?

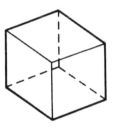

_____

3 How many faces does a cuboid have?

_____

Revision Exercise

1 a Measure these lines and record their lengths in the table:

A ————————————————————

B ————————————————

Line	Measurement
A	cm
B	cm

Complete:

b 13 cm to the nearest 5 cm   = _____ cm

c 39 cm to the nearest 5 cm   = _____ cm

d 28 cm to the nearest 10 cm = _____ cm

e 72 cm to the nearest 10 cm = _____ cm

2 Work these out:

a g	**b** g	**c** g	**d** g	**e**
350	643	127	206	5) 780 g
+ 247	− 280	× 5	× 4	

3 a What is the time shown by each clock?

i    **ii**    **iii**    **iv**

_____   _____   _____   _____

b Draw hands on the clock faces to show the given times:

i 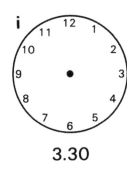   **ii** 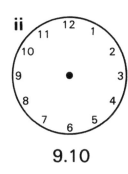   **iii** 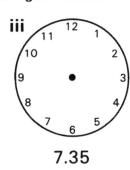   **iv**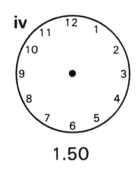

     3.30   9.10   7.35   1.50

4 Write the correct values in the boxes:

Use: [1 ¢]   [5 ¢]   [10 ¢]   [25 ¢]   [$1]   [$2]   [$5]   or   [$10]

a [   ] + [   ] + [   ] = 40 ¢

b [   ] + [   ] + [   ] + [   ] = 56 ¢

c [   ] + [   ] + [   ] + [   ] = 85 ¢

d [   ] + [   ] + [   ] + [   ] = 76 ¢

e ☐ + ☐ + ☐ + ☐ = $2.50

f ☐ + ☐ + ☐ = $13.00

g ☐ + ☐ + ☐ + ☐ = $11.00

h ☐ + ☐ + ☐ + ☐ = $10.35

i ☐ + ☐ + ☐ + ☐ = $14.05

j ☐ + ☐ + ☐ + ☐ = $15.26

5 **a** Devon is 89 cm tall and Akari is 93 cm tall. Who is taller? _____

b Container A holds 800 mℓ of water. Container B holds
1 ℓ of water. Which container holds more? _____

c Alison weighs 37 kg. Her mother weighs twice as much.
How much does Alison's mother weigh? _____

d Mother left home at 8.30 a.m. and arrived at work at
9.30 a.m. How long did she take to get to work? _____

e School begins at 8.45 a.m. André was 30 minutes early.
When did André get to school? _____

f Michael had $10.00. He bought a ball-point pen for
$3.85. How much money did he have left? _____

g What change is left out of $10.00 after buying a litre
of milk for $2.95 and a litre of orange juice for $2.80? _____

h John is 150 cm tall. Jean is 11 cm shorter.
How tall is Jean? _____

5 *Sets*

Exercise 1

1 Sort this set of numbers {36, 35, 21, 23, 5, 6, 40, 28, 27, 38} into odd numbers and even numbers:

   a Odd numbers {_____}

   b Even numbers {_____}

2 From the following units of measurement {gram, litre, minute, kilometre, millilitre, kilogram, hour, centimetre, metre}, list:

   a the set of units measuring length   {_____}

   b the set of units measuring mass   {_____}

   c the set of units measuring time   {_____}

   d the set of units measuring capacity {_____}

3 List the members of the following sets:

   a the set of coins used in your country

      {_____}

   b the set of bank notes used in your country

      {_____}

   c the set of days of the week

      {_____}

   d the set of months of the year

      {_____}

   e the set of counting numbers from 11 to 15

      {_____}

   f the set of odd numbers between 11 and 19

      {_____}

   g the set of even numbers between 20 and 30

      {_____}

   h the set of numbers on a clock face

      {_____}

4 Look at these diagrams and then complete the following:

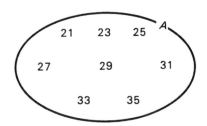

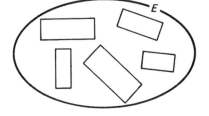

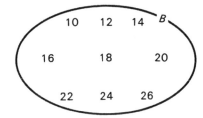

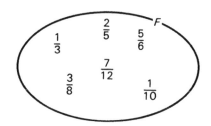

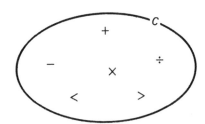

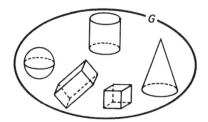

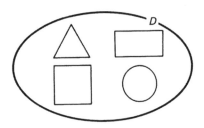

a *A* is a set of _____ with _____ members.

b *B* is a set of _____ with _____ members.

c *C* is a set of _____ with _____ members.

d *D* is a set of _____ with _____ members.

e *E* is a set of _____ with _____ members.

f *F* is a set of _____ with _____ members.

g *G* is a set of _____ with _____ members.

5 Complete to show equal sets:

a

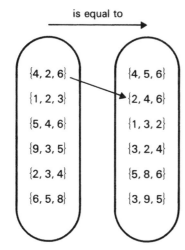

is equal to

{4, 2, 6}   {4, 5, 6}

{1, 2, 3}   {2, 4, 6}

{5, 4, 6}   {1, 3, 2}

{9, 3, 5}   {3, 2, 4}

{2, 3, 4}   {5, 8, 6}

{6, 5, 8}   {3, 9, 5}

b

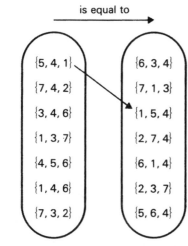

is equal to

{5, 4, 1}   {6, 3, 4}

{7, 4, 2}   {7, 1, 3}

{3, 4, 6}   {1, 5, 4}

{1, 3, 7}   {2, 7, 4}

{4, 5, 6}   {6, 1, 4}

{1, 4, 6}   {2, 3, 7}

{7, 3, 2}   {5, 6, 4}

6   *Graphs*

Exercise 1

The graph shows the number of fruit trees in our school yard:

Mango	🌳 🌳
Orange	🌳 🌳 🌳 🌳 🌳 🌳
Pawpaw	🌳 🌳 🌳 🌳
Pear	🌳
Grapefruit	🌳 🌳 🌳
Lime	🌳 🌳 🌳 🌳 🌳 🌳 🌳 🌳
	🌳 = 1 tree

1 How many orange trees are in our school yard?   _____

Complete:

2 There are twice as many lime trees as _____ trees.

3 There are half as many _____ trees as pawpaw trees.

4 How many more orange trees than mango trees are there? _____

5 How many fruit trees are there in the school yard? _____

116

Exercise 2

The graph shows the number of points scored by the five houses of our school during a spelling competition:

Red house

Yellow house

Blue house

Green house

Orange house

= 2 points

1 Who won the spelling competition? _____

2 How many points did the orange house score? _____

3 Which house scored half as many points as the blue house? _____

4 Which house scored 8 points? _____

5 Which house scored 6 points less than the orange house? _____

Exercise 3

The graph shows the attendance for Class 2 (of which there are 20 pupils):

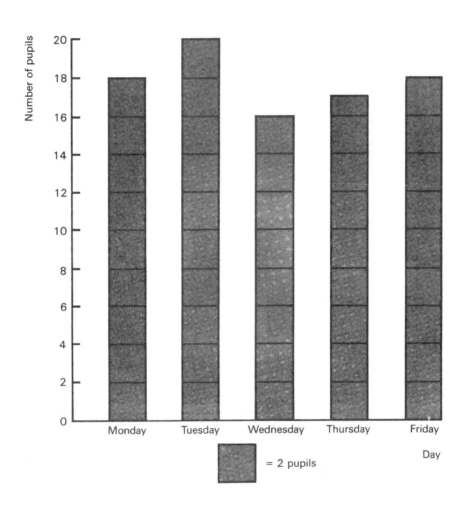

Answer these for Class 2:

1 How many pupils came to school on Thursday?   _____

2 How many pupils were absent on Wednesday?   _____

3 On which day did all the pupils come to school?   _____

4 On which days did the same number of pupils come to school?

_____

Exercise 4

The graph shows the favourite subject for each pupil of a class:

Reading	👤 👤
Art	👤 👤 👤 👤 👤
Mathematics	👤 👤 👤 👤 👤 👤
Grammar	👤 👤 👤 👤 👤 👤 👤 👤 👤 👤
Science	👤 👤 👤 👤
Social studies	👤 👤 👤 👤

👤 = 1 pupil

1 How many pupils like mathematics best? _____

2 Which is the most popular subject among the pupils? _____

3 Which subject is as popular as science? _____

4 How many pupils are in the group? _____

5 How many more pupils like grammar than art? _____

Exercise 5

The graph shows the favourite game for each senior pupil of a certain group:

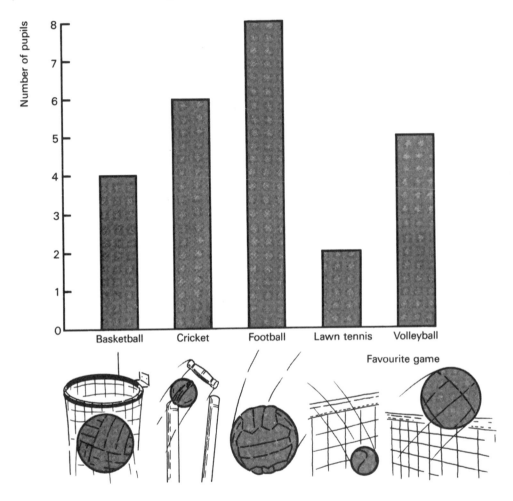

Complete:

1 5 pupils like _____

2 Twice as many pupils like football as _____

3 Which game is most popular among the pupils? _____

4 How many more pupils like football than cricket? _____

5 Which game do the least number of pupils like? _____

6 How many pupils are in the group? _____

Exercise 6

The graph shows the number of hours of sunshine that we had during a certain week in October:

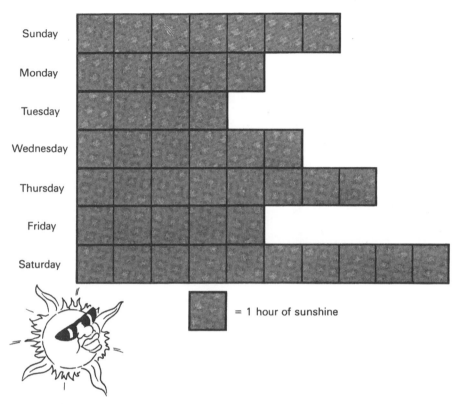

= 1 hour of sunshine

Answer these for that week:

1 How many hours of sunshine were there on Sunday?

_____

2 On which day did the Sun shine for 6 hours? _____

3 Which day had the same number of hours of sunshine as Friday?

_____

4 Which day had the least number of hours of sunshine? _____

5 Which day had twice as many hours of sunshine as Tuesday?

_____

6 How many hours of sunshine in all were there for the week?

_____

Paper 1

Work these out:

1 6 + 4 + 3 = _____

2 7 + 5 + 5 = _____

3 4 + 9 + 1 = _____

4
$$\begin{array}{r} 7 \\ 4 \\ +\ 5 \\ \hline \\ \hline \end{array}$$

5
$$\begin{array}{r} 3 \\ 9 \\ +\ 6 \\ \hline \\ \hline \end{array}$$

6 15 − 9 = _____

7 13 − 4 = _____

8 11 − 8 = _____

9
$$\begin{array}{r} 76 \\ -\ 30 \\ \hline \\ \hline \end{array}$$

10
$$\begin{array}{r} 83 \\ -\ 63 \\ \hline \\ \hline \end{array}$$

11 Shelly had 35 cherries. She gave 10 of them to Ann-Marie. How many cherries did she have left? _____

12 Mark caught 20 fish from the river. Calvin caught 25 fish from the sea. How many fish did they catch altogether? _____

13
$$\begin{array}{r} 23 \\ \times\ 3 \\ \hline \\ \hline \end{array}$$

14
$$\begin{array}{r} 92 \\ \times\ 4 \\ \hline \\ \hline \end{array}$$

15 39
   × 2
   _____

16 116
   × 5
   _____

17 84 ÷ 4 = _____

18 92 ÷ 4 = _____

19 360 ÷ 3 = _____

20 584 ÷ 2 = _____

21 Aunty Pam shared 48 lollipops among her 4 nieces. How many lollipops did each niece receive? _____

22 What is 13 times 3? _____

Write the missing number in each row:

23 5, 10, _____, 20, 25

24 7, 10, 13, 16, _____

25 _____, 10, 20, 30, 40

26 11, 13, 15, 17, _____

27 Debbie, our house captain, is 130 cm tall.
Her sister, Michelle, is 8 cm shorter.
How tall is Michelle? _____

Write +, −, ×, or ÷ in each circle to make each statement true:

28 5 ◯ 3 = 3 ◯ 5

29 8 ◯ 4 = 4 ◯ 3

30 10 ◯ 2 = 4 ◯ 1

31 7 ◯ 3 = 8 ◯ 2

32 6 ◯ 1 = 6 ◯ 0

Write in figures:

33 three hundred and ten   _____

34 two thousand, one hundred and eight _____

35 five thousand and sixteen   _____

36 six thousand and two   _____

Write the missing numbers in the boxes:

37 $5 \times \boxed{} = 20$ **39** $8 + \boxed{} = 13$

38 $\boxed{} - 3 = 9$ **40** $\boxed{} \div 5 = 3$

Paper 2

Work these out:

1	**2**	**3**	**4**
36 + 21	23 14 + 32	410 + 265	242 130 + 315

5 Omar knocked down 12 mangoes, Ryan knocked down 18 mangoes and Dwayne knocked down 25 mangoes. How many mangoes did the three boys knock down altogether?   _____

6	**7**	**8**	**9**
96 − 42	754 − 241	824 − 620	8564 − 2133

10 Shawn's mother gave him 50 bubble-gum pictures. He gave 18 of them to friends, How many bubble-gum pictures did he have left? _____

11   243
   ×   3
_____

12   168
   ×   4
_____

13   207
   ×   4
_____

14   190
   ×   5
_____

15 What is the product of 24 and 4? _____

16   98 ÷ 7 = _____

18 618 ÷ 6 = _____

17 700 ÷ 5 = _____

19 540 ÷ 4 = _____

What is the time shown by each clock?

20

_____

22

_____

21

_____

23

_____

Answer these:

24 5 girls share 60 crayons equally. How many will each girl receive? _____

25 Ruth wants to buy a colouring book for
$2.45 and a packet of markers for $1.95.
How much money does she need?   _____

26 300 + 60 + 4 = _____

27 1000 + 800 + 60 + 5 = _____

28 2000 + 500 + 6 = _____

29 4000 + 10 + 3 = _____

What fraction is shaded and what fraction is not shaded?

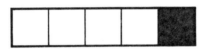

30 Shaded _____

31 Not shaded _____

32 Shaded _____

33 Not shaded _____

34 Shaded _____

35 Not shaded _____

Choose one of these names: circle, triangle, sphere, square, rectangle, to
identify each of these shapes:

36    **37**    **38**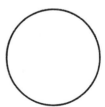

_____   _____   _____

39  _____

40 ☐ _____

Paper 3

Work these out:

| **1** | 243
162
+ 371
_____ | **2** | 4836
+ 2079
_____ | **3** | 7426
− 2390
_____ | **4** | 286
×   3
_____ |

Arrange the numbers in each row in order of size, smallest first:

5 30 43 19 27 36 _____

6 28 12 35 41 21 _____

7 50 37 46 39 42 _____

8 63 71 38 49 53 _____

9 Mr Walkes, the farmer, has 300 pigs while
Mr Griffith has 267 pigs. How many more pigs
does Mr Walkes have than Mr Griffith?   _____

Draw hands on the clock faces to show the given times:

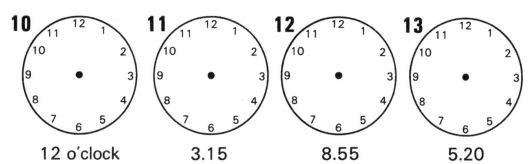

10 12 o'clock   **11** 3.15   **12** 8.55   **13** 5.20

What is the place value of the 6 in each number (thousands, hundreds, tens or ones)?

14 4<u>6</u>3 _____

16 <u>6</u>431 _____

15 2<u>6</u>08 _____

17 957<u>6</u> _____

18 A mini-bus carries 25 passengers on every trip. How many passengers will it carry on 8 trips?   _____

Complete the following for the set of leaves:

19 $\frac{1}{2}$ of them = _____

20 $\frac{1}{4}$ of them = _____

21 $\frac{1}{3}$ of them = _____

22 $\frac{1}{6}$ of them = _____

23 $\frac{1}{12}$ of them = _____

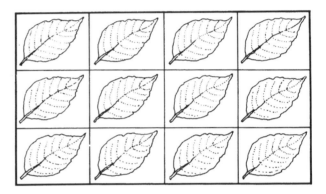

Answer these:

24 32
   × 20
  _____

25 47
   × 20
  _____

26 6⟌102

_____

27 8⟌840

_____

Here are some items sold in our school shop:

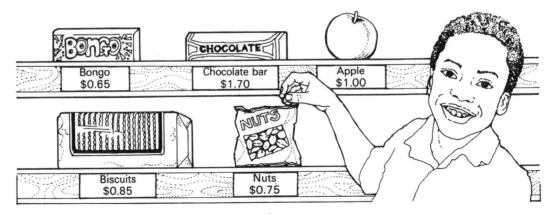

Bongo $0.65   Chocolate bar $1.70   Apple $1.00

Biscuits $0.85   Nuts $0.75

28 Peter bought a packet of nuts and a chocolate bar. How much money did he spend? _____

29 Suzette wants to buy a Bongo and a packet of biscuits. How much money does she need? _____

30 Jamar had $5.00. What change is left after he bought an apple and a packet of nuts? _____

31 Sheena had $10.00. She bought 3 apples and 2 Bongos. How much money did she have left? _____

32 Which item costs half as much as a chocolate bar? _____

Write the missing numbers in the boxes:

33 ☐ + 6 = 15

34 12 − ☐ = 8

35 ☐ ÷ 3 = 8

36 ☐ × 5 = 35

37 ☐ − 5 = 9

38 8 × ☐ = 32

39 27 ÷ ☐ = 9

40 9 + ☐ = 17

Paper 4

Write in figures:

1 seven thousand and eight _____

2 three hundred and forty-two _____

3 two thousand, five hundred and one _____

4 eight thousand and forty _____

5 five thousand and thirty-six _____

Work these out:

6   3246
     + 1507

_____

7   8420
     − 6218

_____

8   207
    ×   4

_____

9 920 ÷ 3 = _____

10   27
     × 30

_____

11   58
     × 30

_____

12   70
     × 30

_____

13   80
     × 30

_____

14   96
     × 30

_____

The graph shows the points scored by the five houses of our school during the sports and games competition:

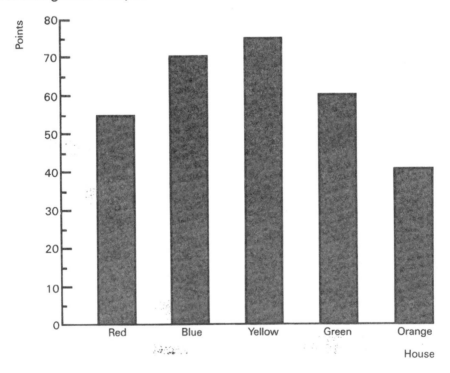

15 Which house won the competition? _____

16 How many points did the blue house score? _____

17 How many points did the house which came last score? _____

18 How many houses scored more points than the green house? _____

19 Which house was placed second in the competition? _____

Work these out:

20 $\frac{2}{5} + \frac{1}{5} =$ _____   **23** $\frac{1}{6} + \frac{1}{6} + \frac{5}{6} =$ _____

21 $\frac{7}{8} - \frac{3}{8} =$ _____   **24** $\frac{9}{10} - \frac{3}{10} =$ _____

22 $\frac{7}{12} + \frac{5}{12} =$ _____   **25** $\frac{5}{2} - \frac{3}{2} =$ _____

26 Our school concert began at 2.30 p.m. and finished at 6.00 p.m. How long did it last? _____

27 How many times can I subtract 7 from 70? _____

28 Aunty Pat made 60 sugar cakes. How many members of the family could get 5 sugar cakes each? _____

In each line there is an even number. Circle it.

29 6   11   9   15   21   **32** 32   25   41   37   43

30 31   27   13   10   7   **33** 39   53   48   69   33

31 29   19   5   17   24

34 Kemar is 17 years old. His father is 3 times as old, how old is Kemar's father? _____

35 Kim weighs 39 kg. Kelly weighs 15 kg more than Kim. What is Kelly's weight? _____

Write the missing numbers in the boxes:

36 13 − 5 = 4 × ☐   **39** ☐ − 3 = 9 − 1

37 4 × 8 = 8 × ☐   **40** 5 × ☐ = 10 + 10

38 6 × 0 = 5 − ☐

Paper 5

What is the time shown by each clock?

1    **2**    **3**    **4**

_____   _____   _____   _____

Answer these:

5 Our reading lesson began at 11.45 a.m.
and finished 30 minutes later.
At what time did it finish?   _____

6   3497
     + 2305
     _____

     _____

7   6520
     − 3158
     _____

     _____

8   168
      ×   7
      _____

      _____

9 504 ÷ 4 = _____

Write the following measurements to the nearest 5 cm:

10 16 cm = _____   **11** 21 cm = _____

12 9 cm = _____   **14** 43 cm = _____

13 38 cm = _____   **15** 52 cm = _____

Multiply these:

16 21   **17** 26   **18** 75   **19** 89
$\times$ 40 $\times$ 40 $\times$ 40 $\times$ 40

_____   _____   _____   _____

In the number 4065, which digit stands for the following?

20 hundreds _____   **22** ones   _____

21 tens   _____   **23** thousands _____

Work these out:

24 225 pencils are put in packets of 5.
How many packets will there be?   _____

25 My mother weighs 73 kg. My father
weights 69 kg. Who is lighter?   _____

26 After buying a drawing book for $1.35,
Ann had $3.65 left. How much money did
she have at first?   _____

Choose one of these names: cone, circle, triangle, square, cube, rectangle, to
identify each of these shapes:

27   **29**    **31**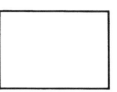

_____   _____   _____

28    **30** 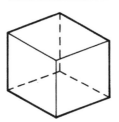   **32**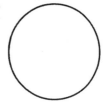

_____   _____   _____

Colour:

33 $\frac{1}{2}$ of this set   △ △ △ △ △ △

34 $\frac{1}{4}$ of this set   ☐ ☐ ☐ ☐ ☐ ☐ ☐ ☐

35 $\frac{1}{5}$ of this set   ○ ○ ○ ○ ○ ○ ○ ○ ○ ○

36 $\frac{1}{3}$ of this set   ▭ ▭ ▭ ▭ ▭ ▭

Write the missing numbers in the boxes:

37 7 + 6 = 6 + ☐   **39** 8 × ☐ = 0

38 9 × ☐ = 9   **40** 5 + 0 = ☐

Paper 6

In each row there is an odd number, circle it:

1 12   6   15   14   20   **4** 38   13   52   10   22

2 17   18   24   26   8   **5** 62   54   36   39   40

3 16   28   30   32   21

Measure the following lines and write the answers in centimetres:

6 _____   _____

7 _____   _____

8 _____   _____

9 _____   _____

10 On Monday, I helped my mother pick 860 limes; on Tuesday, we picked 1085 limes. How many limes did we pick in all? _____

Use one of these words: thousands, hundreds, tens, ones to complete the following:

11 257 = 2 _____ + 5 _____ + 7 _____

12 3406 = 3 _____ + 4 _____ + 0 ____ + 6 ____

13 5723 = 5 _____ + 7 _____ + 2 ____ + 3 ____

14 6800 = 6 _____ + 8 _____ + 0 ____ + 0 ____

Answer these:

15 Tricia sold 180 bunches of lettuce to the supermarket, Monica sold 225 bunches of lettuce to a hotel. How many more bunches of lettuce did Monica sell than Tricia? _____

16 What number multiplied by 6 gives 84? _____

17
```
  347
  260
+ 174
_____

_____
```

20 875 ÷ 7 = _____

23
```
    38
 × 50
_____

_____
```

18
```
  8420
- 3156
_____

_____
```

21
```
   61
 × 50
_____

_____
```

24
```
   74
 × 50
_____

_____
```

19
```
  107
 ×  8
_____

_____
```

22
```
   29
 × 50
_____

_____
```

Here are four containers of milk with their measurements in millilitres:

A   **B**   **C**   **D**

25 Which container holds the most milk?   _____

26 Which container holds half as much milk as C? _____

27 How many times can I fill container A from D? _____

28 Which container holds twice as much as B?   _____

29 One litre is 1000 mℓ. How many millilitres short
of a litre is container D?   _____

Write +, −, ×, or ÷ in each circle to make each statement true:

30 5 ◯ 2 = 9 ◯ 2   **33** 4 ◯ 1 = 4 ◯ 0

31 8 ◯ 4 = 2 ◯ 6   **34** 16 ◯ 8 = 1 ◯ 2

32 0 ◯ 5 = 7 ◯ 7   **35** 9 ◯ 3 = 3 ◯ 2

Write in words:

36 4615 _____

37 1073 _____

38 3007 _____

39 480 _____

40 5600 _____

Paper 7

Draw an arrow from *A* to *B* to show 'is equal to':

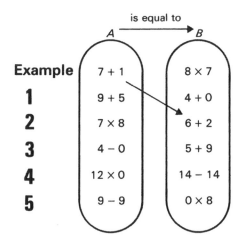

Write the following measurements to the nearest 10 cm:

6 8 cm = _____ cm   **9** 41 cm = _____ cm

7 23 cm = _____ cm   **10** 57 cm = _____ cm

8 36 cm = _____ cm   **11** 34 cm = _____ cm

Work these out:

12   425
         1613
       + 2730
       _____

       _____

13   9385
       − 2760
       _____

       _____

14   167
        ×   8
        _____

        _____

15 875 ÷ 7 = ____

16 Antonio has 50 marbles. There are 16 red marbles and 15 blue marbles; the other marbles are green. How many green marbles does Antonio have?   _____

17 Half of André's money is $2.75.
How much money does he have? _____

18   7
     × 70
     _____

     _____

19   26
     × 70
     _____

     _____

20   39
     × 70
     _____

     _____

21 What number added to 25 gives 42? _____

The table shows three sets of marks, each out of 25, for six girls of Class 2:

Pupil	Shekira	Isha	Rena	Ayesha	Vena	Keisha
Grammar mark	21	19	25	20	21	16
Mathematics mark	24	20	15	16	19	23
Science mark	18	17	20	22	17	19
Total mark						

22 Who scored the highest mark in mathematics?   _____

23 Who scored the lowest mark in grammar?   _____

24 Which two girls scored the same marks in science?   _____

Work out the total mark for each girl and complete the table above; then answer these:

25 Who scored the highest total mark?   _____

26 Which two girls scored the same total marks?   _____

27 Who scored the lowest total mark? _____

28 Whose total mark is 3 less than
Rena's?   _____

Write the missing numbers in the boxes:

29 ☐ + 3 = 12   **32** ☐ × 6 = 48

30 15 − ☐ = 8   **33** 32 ÷ ☐ = 4

31 ☐ − 4 = 9   **34** ☐ ÷ 3 = 9

Use the diagram to help you answer these:

35 What is $\frac{1}{4}$ of 20? _____

36 What is $\frac{1}{5}$ of 20? _____

Today is Sunday, 16th October:

37 What date was last Sunday?   _____

38 What date will it be next Sunday? _____

Here are four digits: ② ⑨ ③ ⑦
Use each digit once only to form:

39 the largest possible number _____

40 the smallest possible number _____

Paper 8

Use the diagram below to help you answer the following questions:

1 WHOLE							
$\frac{1}{2}$				$\frac{1}{2}$			
$\frac{1}{4}$		$\frac{1}{4}$		$\frac{1}{4}$		$\frac{1}{4}$	
$\frac{1}{8}$	$\frac{1}{8}$	$\frac{1}{8}$	$\frac{1}{8}$	$\frac{1}{8}$	$\frac{1}{8}$	$\frac{1}{8}$	$\frac{1}{8}$

1 $\frac{3}{8}+\frac{4}{8}=$ _____

2 $\frac{1}{8}+\frac{1}{8}+\frac{3}{8}=$ _____

3 $\frac{7}{8}-\frac{5}{8}=$ _____

4 $\frac{3}{4}+\frac{3}{4}=$ _____

5 $1-\frac{3}{4}=$ _____

6 $1-\frac{5}{8}=$ _____

7 $1-\frac{7}{8}=$ _____

8 $1-\frac{1}{2}=$ _____

Suzette has 5 green markers and 3 red markers:

9 What fraction of the markers are green? _____

10 What fraction of the markers are red? _____

Write the answer for:

11 $2000+300+50+1=$ _____

12 $5000+100+60+3=$ _____

13 $1000+700+9=$ _____

14 $6000+40+5=$ _____

15 There are 30 pupils in Class 2; 18 of the
pupils are girls. How many boys are in
Class 2?   _____

16   9   **17**   18   **18**   27   **19**   46
     × 80   × 80   × 80   × 80

_____   _____   _____   _____

20 Grandad shared 75 golden apples equally among
his 5 grandchildren. How many apples did each
child receive?   _____

Here are the weights of five boys from Class 2:

Pedro   André   Mario   Akari   Carlo
40 kg   38 kg   36 kg   45 kg   35 kg

21 Who is the heaviest?   _____

22 Who is the lightest?   _____

23 How much heavier is Akari than
Mario?   _____

24 Who is 10 kg lighter than Akari? _____

25 What is the total weight of André
and Mario?   _____

Work these out:

26   3178
        + 4636
        _____

        _____

27   8624
        − 5170
        _____

        _____

28   609
        ×   9
        _____

        _____

29 936 ÷ 9 = _____

Write the missing number in each sequence:

30 4, 9, 14, 19, _____

31 25, 23, 21, 19, _____

32 12, 23, 34, 45, _____

33 _____, 12, 18, 24, 30

What is the value of each digit in the number, 4096 (thousands, hundreds, tens or ones)?

34 9 = 9 _____

35 0 = 0 _____

36 4 = 4 _____

37 6 = 6 _____

Sandra had $10.00. She bought a packet of crayons for $2.95 and a colouring book for $2.25.

38 How much money did she spend?   _____

39 How much money did she have left? _____

Here is a cuboid: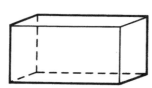

40 How many faces does it have? _____

Paper 9

Write in figures:

1 two thousand, five hundred and sixty-two _____

2 seven thousand two hundred _____

3 one hundred and eighty-seven _____

4 three thousand and nine _____

5 five thousand and ten _____

Work these out:

6
```
  2461
   104
+ 3275
_____
```

7
```
  8134
- 2076
_____
```

8
```
   480
×    8
_____
```

9 8) 840
_____

How much time has passed? (Each pair of clocks shows times in the same day.)

10 from   to
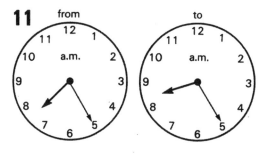
_____

12 from   to

_____

11 from   to
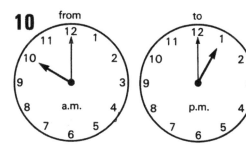
_____

13 from   to

_____

Today is Wednesday, 16th March:

14 What date was Wednesday last
week?   _____

15 What date will it be next
Wednesday?   _____

Work these out:

16   8   **17**   27   **18**   36   **19**   59
      $\times$ 80 $\times$ 80 $\times$ 80 $\times$ 80

      _____   _____   _____   _____

20 6 bags of potatoes weigh 102 kg. If each bag
contains the same amount, what is the weight of
each bag?   _____

The graph shows the different types of snack the pupils from Class 2 buy during
lunch-time one day:

Corncurls	ⵣ ⵣ ⵣ ⵣ ⵣ ⵣ ⵣ ⵣ ⵣ ⵣ
Zoomers	ⵣ ⵣ ⵣ ⵣ
Cheesezees	ⵣ ⵣ ⵣ ⵣ ⵣ ⵣ
Popcorn	ⵣ ⵣ ⵣ ⵣ ⵣ
Sweet biscuits	ⵣ ⵣ
Bongos	ⵣ ⵣ ⵣ
	ⵣ = 1 pupil

21 Which snack was most popular among the pupils? _____

22 Which snack did the least number of pupils buy? _____

23 Which snack was half as popular as Corncurls? _____

24 How many pupils bought Zoomers? _____

25 Which snack was twice as popular as Bongos? _____

26 How many more pupils bought Corncurls than Cheesezees? _____

27 How many pupils bought snacks? _____

Look at this set of numbers {11, 9, 8, 15, 14, 18, 19, 24}.
From this set, write:

28 the smallest odd number _____

29 the largest even number _____

30 the largest odd number _____

31 the smallest even number _____

32 the number which can be divided exactly by 5 _____

Complete the following for the set of eggs:

33 $\frac{1}{2}$ of them = _____

34 $\frac{1}{3}$ of them = _____

35 $\frac{1}{4}$ of them = _____

36 $\frac{1}{6}$ of them = _____

37 $\frac{1}{8}$ of them = _____

38 $\frac{1}{12}$ of them = _____

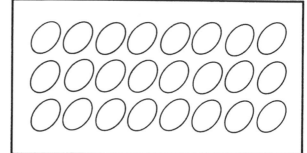

A teacher has 7 blue markers and 5 green markers:

39 What fraction of the markers are blue? _____

40 What fraction of the markers are green? _____

Paper 10

Write the missing number in each row:

1 3, 8, 13, 18, _____

2 40, 36, 32, 28, _____

3 _____, 14, 21, 28, 35

4 8, 16, _____, 32, 40

Answer these:

5
```
   357
    23
 + 416
 ─────

 ─────
```

7
```
   278
 ×   6
 ─────
```

6
```
  6520
 -1316
 ─────

 ─────
```

8 $804 \div 6 =$ _____

9 How many times can 6 be taken from 42?   _____

10 What is the remainder when 50 is divided by 6? _____

Complete the following:

11 2176 = __ thousands + __ hundred + __ tens + __ ones

12 3054 = __ thousands + __ hundreds + __ tens + __ ones

13 5003 = __ thousands + __ hundreds + __ tens + __ ones

14 297 = __ hundreds + __ tens + __ ones

Use the diagram to help you complete the following:

1 WHOLE											
$\frac{1}{2}$						$\frac{1}{2}$					
$\frac{1}{3}$				$\frac{1}{3}$				$\frac{1}{3}$			
$\frac{1}{4}$			$\frac{1}{4}$			$\frac{1}{4}$			$\frac{1}{4}$		
$\frac{1}{6}$		$\frac{1}{6}$		$\frac{1}{6}$		$\frac{1}{6}$		$\frac{1}{6}$		$\frac{1}{6}$	
$\frac{1}{12}$	$\frac{1}{12}$	$\frac{1}{12}$	$\frac{1}{12}$	$\frac{1}{12}$	$\frac{1}{12}$	$\frac{1}{12}$	$\frac{1}{12}$	$\frac{1}{12}$	$\frac{1}{12}$	$\frac{1}{12}$	$\frac{1}{12}$

15 $\frac{1}{2} = \frac{}{4}$

16 $\frac{1}{3} = \frac{}{6}$

17 $\frac{1}{2} = \frac{}{12}$

18 $\frac{1}{4} = \frac{}{12}$

19 $\frac{2}{3} = \frac{}{6}$

20 $\frac{2}{3} = \frac{}{12}$

21 $\frac{5}{6} = \frac{}{12}$

22 $\frac{3}{4} = \frac{}{12}$

Here is a cube:

23 What is the shape of each face?

_____

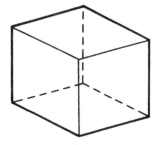

Draw an arrow from *L* to *M* to show 'is equal to':

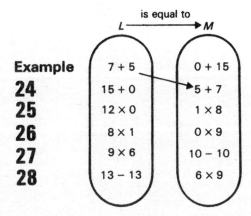

	is equal to
	$L \longrightarrow M$
Example	7 + 5 → 0 + 15
24	15 + 0 → 5 + 7
25	12 × 0 → 1 × 8
26	8 × 1 → 0 × 9
27	9 × 6 → 10 − 10
28	13 − 13 → 6 × 9

What is the time shown by each clock?

  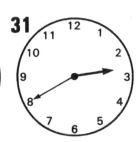 

29 _____ **30** _____ **31** _____ **32** _____

33 A concert began at 2.15 p.m. and was half an
hour late. At what time should it have begun? _____

Write +, −, ×, or ÷ in each circle to make each statement true:

34 7 ◯ 2 = 9 ◯ 5

35 2 ◯ 9 = 3 ◯ 6

36 16 ◯ 2 = 5 ◯ 3

37 15 ◯ 7 = 10 ◯ 2

38 7 ◯ 9 = 4 ◯ 4

39 10 ◯ 10 = 18 ◯ 0

A tin of evaporated milk is a good example of a cylinder
closed at both ends:

40 How many faces does this cylinder
have? _____

Paper 11

Work these out:

1	863 + 124	4	8210 − 3156	7 $636 \div 6 =$ _____

2	3476 + 2735	5	57 × 10	8 $504 \div 7 =$ _____

3	6492 − 1370	6	64 × 30	

In each row, find the biggest number and say whether it is odd or even:

Example

15, 27, 36, 19, 29   *36*   *even*

9 22, 16, 24, 32, 28 _____ _____

10 31, 41, 23, 17, 35 _____ _____

11 40, 25, 33, 21, 38 _____ _____

12 36, 48, 57, 43, 54 _____ _____

13 39, 62, 59, 49, 47 _____ _____

Complete:

14 $\frac{1}{2}$ of 16 = _____   16 $\frac{1}{4}$ of 20 = _____

15 $\frac{1}{3}$ of 18 = _____   17 $\frac{1}{5}$ of 20 = _____

The table below shows the number of gold, silver and bronze medals won by the five houses of our school:

Medal	Gold	4	3	9	6	10
	Silver	7	6	7	8	5
	Bronze	4	5	6	10	8
	Total					
		Blue	Green	Orange	Red	Yellow
				House		

18 Which house won the most gold medals?   _____

19 Which house won the most silver medals?   _____

Work out the total number of medals for each house and complete the table above; then answer these:

20 Which house won the most medals?   _____

21 Which house won the fewest meals?   _____

22 How many medals did the orange house win?   _____

23 If a gold medal is worth 3 points, a silver worth 2 points and a bronze worth 1 point, which house received the most points?   _____

Write the missing numbers in the boxes:

24 $6 \times 3 = 10 + \boxed{}$

27 $\boxed{} + 8 = 3 \times 4$

25 $7 \times 6 = 6 \times \boxed{}$

28 $15 - \boxed{} = 9 - 0$

26 $20 \div 4 = \boxed{} - 4$

29 $16 \div \boxed{} = 8 \times 1$

Here is Pedro's home and the distances he lives from the school, library and the church:

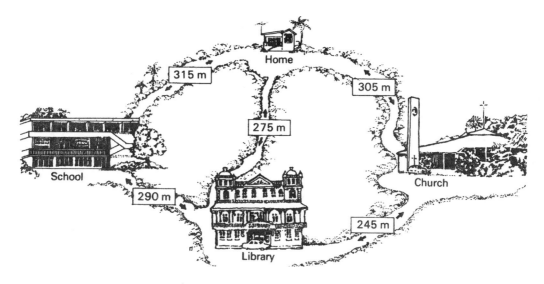

30 Which one of the school, library or church is nearest to Pedro's home? _____

31 Pedro went to school and then went to the library. How far did he travel? _____

32 How far is it from the school to the church passing by way of the library? _____

33 How much further is it from Pedro's home to the school than to the church? _____

Here are four digits: ① ⑤ ④ ⑦
Use each digit once only to form:

34 the smallest possible number _____

35 the largest possible number _____

Write the answers for these:

36 Find the sum of 37, 529 and 83.   _____

37 What is the difference between 200 and 137? _____

38 What number is 15 less than 100?   _____

39 What number divided by 6 gives 18?   _____

40 Which holds more, a container
holding 1ℓ or a container holding
750 mℓ?   _____

Paper 12

Add these:

1	61	**2**	243	**3**	6140	**4**	59
	24		102		+ 1543		27
	+ 13		+ 233				+ 17

Write in figures:

5 five thousand and fourteen   _____

6 seven hundred and twenty   _____

7 three thousand, one hundred and eight _____

8 two thousand and thirty-seven   _____

There are 12 biscuits in one packet:

9 How many biscuits are there in 4 packets? _____

Subtract these:

10	94 − 20	**11**	65 − 15	**12**	398 − 72	**13**	8476 − 2185
	————		————		————		————

What fraction of this set is shaded and what fraction is not shaded?

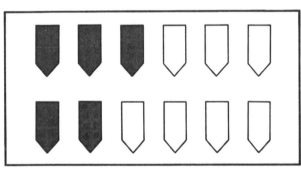

14 Shaded _____

15 Not shaded _____

In a mathematics test, Ayanna got 7 questions correct and 3 wrong:

16 What fraction of the questions did she get correct? _____

17 What fraction of the questions did she get wrong? _____

Multiply these:

18	92 × 4	**19**	19 × 5	**20**	409 × 7	**21**	237 × 8
	————		————		————		————

22 How many pupils can get 5
pencils each from a box containing
120 pencils?   _____

Divide these:

23 90 ÷ 6 = _____   **25** 144 ÷ 8 = _____

24 570 ÷ 5 = _____   **26** 420 ÷ 4 = _____

27 At the zoo, there were 18 big
elephants and 5 small elephants.
How many elephants were there
in all?   _____

Write the answers in the diagram:

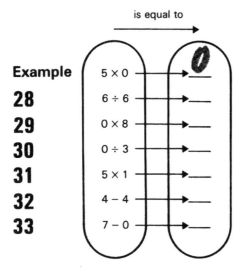

is equal to

Example   5 × 0 ──→ **0**
28   6 ÷ 6 ──→ ___
29   0 × 8 ──→ ___
30   0 ÷ 3 ──→ ___
31   5 × 1 ──→ ___
32   4 − 4 ──→ ___
33   7 − 0 ──→ ___

Here are three digits: (3) (5) (7)

We can form six different three-digit numbers by using each digit once only.
Form them:

34 _____   **37** _____

35 _____   **38** _____

36 _____   **39** _____

40 Grey saved $8.50 and Kevin saved $6.75.
How much more money did Grey save than Kevin? _____

Paper 13

Write the following numbers in words:

1 4316 _____

2 2076 _____

3 1000 _____

4 5300 _____

Each of the following clocks is 30 minutes fast. Write the correct times.

5    **6**    **7**    **8**

_____ _____ _____ _____

Work these out:

9   4678
     + 2892
_____

11   370
     ×   7
_____

10   5780
     − 1873
_____

12 576 ÷ 8 = _____

Ben is 157 cm tall and Carl is 163 cm tall:

13 Who is taller? _____

14 By how much is he taller? _____

Multiply these:

15	87	**16**	69	**17**	12	**18**	23
	× 10		× 20		× 80		× 70

Answer these:

19 Mr Smith caught 600 flying fish
on Saturday and 875 on Sunday.
How many flying fish did he catch
for the two days?   _____

20 Mother made 19 fish cakes and
she gave 4 away. How many fish
cakes did she have left?   _____

For my birthday, my mother bought me the following items:

Reading Book $4.75

Crayons $2.95

Colouring Book $2.25

21 Which was the most expensive
item?   _____

22 Which was the least expensive
item?   _____

23 Which two items add up to $7.00?
_____

24 How much money did she spend
in all?   _____

25 Mother had $20.00. How much
money does she have left after
buying the three items?   _____

Use the diagram below to help you put the correct signs, <, = or >, in the circles and also to find the answers to questions 32–37:

1 WHOLE							
$\frac{1}{2}$				$\frac{1}{2}$			
$\frac{1}{4}$		$\frac{1}{4}$		$\frac{1}{4}$		$\frac{1}{4}$	
$\frac{1}{8}$	$\frac{1}{8}$	$\frac{1}{8}$	$\frac{1}{8}$	$\frac{1}{8}$	$\frac{1}{8}$	$\frac{1}{8}$	$\frac{1}{8}$

26 $\frac{1}{2}$ ◯ $\frac{1}{4}$

27 $\frac{1}{4}$ ◯ $\frac{2}{8}$

28 $\frac{3}{4}$ ◯ $\frac{5}{8}$

29 $\frac{3}{8}$ ◯ $\frac{3}{4}$

30 $\frac{4}{8}$ ◯ $\frac{1}{2}$

31 $\frac{7}{8}$ ◯ $\frac{3}{4}$

32 $\frac{1}{4} + \frac{1}{4} + \frac{1}{4} =$ _____

33 $\frac{7}{8} - \frac{5}{8} =$ _____

34 $\frac{1}{2} + \frac{1}{2} =$ _____

35 $1 - \frac{3}{8} =$ _____

36 $1 - \frac{1}{4} =$ _____

37 $1 - \frac{1}{2} =$ _____

Susan had 10 guavas; she ate $\frac{1}{5}$ of them:

38 How many guavas did she eat? _____

39 How many guavas did she have left? _____

40 The gardener had 18 flower pots. He planted 4 seeds in each flower pot. How many seeds did he plant? _____

Paper 14

Write the missing number in each row:

1 30, 27, 24, 21, _____

2 _____, 35, 37, 39, 41

3 21, 31, _____, 51, 61

4 1, 7, 13, 19, _____

Answer these:

5 5 friends shared 125 sweets equally. How many sweets did each friend get?   _____

6 Steve picked 320 limes, Mark picked 186 limes and Joe picked 275 limes. How many limes did they pick in all?   _____

The graph shows the attendance for Class 2 for one week. There are 24 pupils in Class 2.

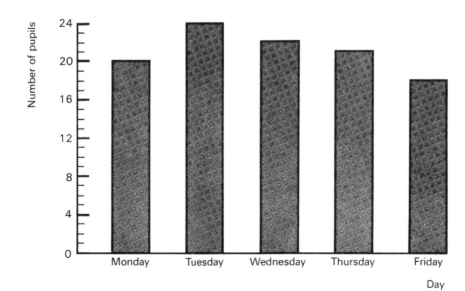

7 How many pupils came to school on Monday? _____

8 On which day did all the pupils come to school? _____

9 How many pupils were absent on Wednesday? _____

10 On which day were 6 pupils absent? _____

11 What was the total attendance for the 5 days? _____

Write the following measurements to the nearest 5 cm:

12 17 cm = _____ cm   **15** 38 cm = _____ cm

13 22 cm = _____ cm   **16** 53 cm = _____ cm

14 46 cm = _____ cm   **17** 34 cm = _____ cm

Work these out:

18 $395 \div 5 =$ _____

20
$$\begin{array}{r} 607 \\ \times \quad 8 \\ \hline \\ \hline \end{array}$$

19 $636 \div 6 =$ _____

21
$$\begin{array}{r} 780 \\ \times \quad 7 \\ \hline \\ \hline \end{array}$$

I have 60 plums:

22 How many children can receive 8 plums each? _____

23 How many plums will be left over? _____

Multiply these:

24	37	**25**	68	**26**	92	**27**	41
	$\times$ 30		$\times$ 40		$\times$ 50		$\times$ 90

Write the answer for:

28 $3000 + 500 + 20 + 9 = $ _____

29 $5000 + 100 + 80 + 4 = $ _____

30 $7000 + 80 = $ _____

31 $6000 + 3 = $ _____

32 What change is left from $5.00 after buying a
pencil case for $1.85?   _____

33 Find the total cost of 4 pencils at 15 ¢ each.   _____

34 What number is 10 less than 97?   _____

Write the missing numbers in the boxes:

35 $5 + 7 = 2 \times \boxed{}$

36 $9 - 3 = 6 + \boxed{}$

37 $\boxed{} - 4 = 12 \div 4$

38 $8 \times \boxed{} = 4 \times 8$

39 $10 - 0 = 10 + \boxed{}$

40 $12 - \boxed{} = 7 - 7$

Paper 15

Answer these:

1   287
      ×   6
    _____

2 713 ÷ 4 = _____

3   6210
       − 1704
     _____

4   97
       × 40
     _____

Each clock is 30 minutes slow, write the correct time:

5    **6**    **7**    **8**

_____   _____   _____   _____

9 A football match began at 4.30 p.m.
    and finished at 6.00 p.m.
    How long did it last?   _____

Here are four digits: (4)
Use each digit once only to form:

10 the largest possible four-digit number   _____

11 the smallest possible four-digit number _____

André weighs 31 kg and Akari weighs 27 kg:

12 Who is heavier?   _____

13 By how much is he heavier? _____

Here are three items
that are sold at
our school shop:

Hamburger $1.80   Hot Dog $1.50   Roti $2.25

Mario bought a Roti and a hot dog; Ayanna bought a Roti and a hamburger; Shakira
bought a hot dog and a hamburger:

14 Who spent the most money?   _____

15 Who spent the least money?   _____

16 Mario had $5.00. How much
money did he have left?   _____

17 Ayanna had $10.00. How much
money did she have left?   _____

18 Shakira had $10.00. How much
money did she have left?   _____

Choose one of these names: cone, circle, sphere, cube, triangle, rectangle,
square, cuboid, to identify each of these shapes:

19   **22**

_____   _____

20   **23**

_____   _____

21   **24**

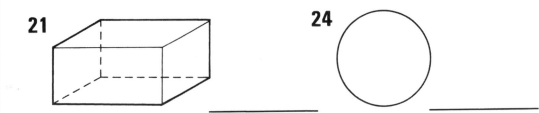

_____   _____

Write the missing numbers in the boxes:

25 ☐ − 6 = 9   28 8 × ☐ = 40

26 13 − ☐ = 5   29 10 + ☐ = 15

27 ☐ ÷ 6 = 8   30 42 ÷ ☐ = 6

Write the missing number in each row:

31 _____, 15, 25, 35, 45

32 7, 10, 13, 16, _____

33 8, 12, _____, 20, 24

34 49, 44, 39, 34, _____

In the number 7539, write which digit stands for:

35 tens   _____   37 ones   _____

36 thousands _____   38 hundreds _____

Carlo has 5 white pigeons and 3 black pigeons:

39 What fraction of the pigeons are white? _____

40 What fraction of the pigeons are black? _____

8 Two Tests

Test 1

Write in figures:

1 ninety-four   _____

2 five thousand and ten   _____

3 two thousand and six   _____

4 three hundred and eighty-four _____

Work these out:

5 $\begin{array}{r} 87 \\ -\ 26 \\ \hline \end{array}$

8 $525 \div 5 = $ _____

11 $\begin{array}{r} 83 \\ \times\ 10 \\ \hline \end{array}$

6 $\begin{array}{r} 276 \\ +\ 109 \\ \hline \end{array}$

9 $\begin{array}{r} 420 \\ -\ 316 \\ \hline \end{array}$

12 $\begin{array}{r} 97 \\ \times\ 60 \\ \hline \end{array}$

7 $\begin{array}{r} 79 \\ \times\ 4 \\ \hline \end{array}$

10 $\begin{array}{r} 2475 \\ +\ 1709 \\ \hline \end{array}$

13 Find the sum of 67, 329 and 86. _____

14 What number is 20 less than 95? _____

15 Arlette has 75 cherries. To how many friends can she give 5 cherries each? _____

16 How many faces has a cuboid? _____

Write the missing numbers in the boxes:

17 ☐ − 4 = 8

20 9 × ☐ = 45

18 17 − ☐ = 9

21 42 ÷ ☐ = 7

19 ☐ ÷ 6 = 6

22 ☐ + 7 = 15

Answer these:

23 How many times can I take 8 from 64? _____

24 What is the difference between 87 and 200? _____

Write <, = or > in the circles to make each statement true:

25 8 + 4 ◯ 15 − 6

27 6 × 0 ◯ 7 − 7

26 7 × 3 ◯ 3 × 7

28 1 × 5 ◯ 15 ÷ 5

Answer these:

29 A bag of potatoes weighs 45 kg and a bag of onions weighs 38 kg. Which weighs more? _____

30 Mary had 60 mangoes, how many bags are needed to put 5 mangoes in each of them? _____

31 What is the product of 14 and 7? _____

Write the missing number in each row:

32 1, 5, 9, 13, _____

33 36, 31, 26, 21, _____

34 _____, 7, 10, 13, 16

35 7, 14, _____, 28, 35

Write the missing numbers in the boxes:

36   3   ×   6   =   9   +   ☐

37   ☐   ÷   4   =   4   +   0

38   5   ×   1   =   ☐   −   3

39   0   ×   7   =   8   −   ☐

Answer these:

40 Susie wants to buy a reading book for $7.30,
but she has $4.75 only.
How much more money does she need?   _____

41 Ronnie is 147 cm tall, Ronnie's father is
16 cm taller than him. How tall is Ronnie's father?   _____

Write in words:

42   436 = _____

43 1708 = _____

44 5006 = _____

45 2027 = _____

Write the answer for:

46 3000 + 500 + 80 + 6 = _____

47 8000 + 20 + 3 =   _____

48 1000 + 400 + 7 =   _____

In the number 3052, which digit stands for the following?

49 tens _____   **51** ones _____

50 thousands _____   **52** hundreds _____

What is the time shown by each clock?

53

_____

55

_____

54

_____

56

_____

Use the diagram to answer these:

57 What fraction is shaded? _____

58 What fraction is not shaded? _____

59 What is $\frac{1}{2}$ of 10? _____

60 What is $\frac{1}{5}$ of 10? _____

Test 2

Work these out:

1 740 ÷ 4 = _____

3   6320
      − 247
      _____

      _____

2   608
       ×   7
       _____

       .

4   86
       × 40
       _____

Complete these using the words thousands, hundreds, tens or ones:

5   472 = 4 _____ + 7 _____ + 2 _____

6 2056 = 2 _____ + 0 _____ + 5 _____ + 6 _____

7 7380 = 7 _____ + 3 _____ + 8 _____ + 0 _____

Angela had $10.00. She bought a hamburger for $1.80 and a Roti for $2.25.

8 How much money did she spend?   _____

9 How much money did she have left? _____

Answer these:

10 What number when multiplied by 6 gives 24?   _____

11 When a number is divided by 4 the answer is 8, what is
the number?   _____

The table shows the marks for three subjects, each one out of 20, for six boys:

Science	17	15	13	19	16	15
Mathematics	20	12	19	18	14	13
Social studies	19	18	17	14	18	16
Total						
Pupil	Ron	Mario	Reno	Pedro	Akari	André

12 Who scored the highest mark in science? _____

13 Who scored the lowest mark in social studies? _____

Work out the total number of marks for each boy and complete the table above; then answer these:

14 Who scored the highest total mark? _____

15 Who scored full marks in mathematics? _____

16 Who scored the lowest total mark? _____

Draw hands on the clock faces to show the given times:

17 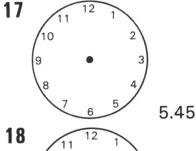 5.45

19 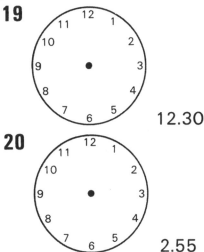 12.30

18 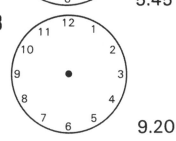 9.20

20 2.55

Answer these:

21 Our mathematics lesson began at 10.15 a.m. and lasted for 30 minutes. At what time did it finish? _____

22 What number added to 37 gives 60? _____

What is the place value of the 6 in each number (thousands, hundreds, tens or ones)?

23 4615 _____   **25** 6157 _____

24 2706 _____   **26** 3860 _____

Write the following measurements to the nearest 10 cm:

27 27 cm = _____ cm   **29** 51 cm = _____ cm

28 44 cm = _____ cm   **30** 39 cm = _____ cm

31 What number when subtracted from 80 gives 27? _____

Complete:

32 873 = __ hundreds + __ tens + __ ones

33 4082 = __ thousands + __ hundreds + __ tens + __ ones

Work these out:

34 Sally is 18 years old. Sally's mother is 3 times as old as her. How old is Sally's mother? _____

35 Half of Mark's money is $2.86. How much money does Mark have? _____

Use the diagram to help you answer the questions that follow:

1 WHOLE									
$\frac{1}{2}$					$\frac{1}{2}$				
$\frac{1}{5}$		$\frac{1}{5}$		$\frac{1}{5}$		$\frac{1}{5}$		$\frac{1}{5}$	
$\frac{1}{10}$	$\frac{1}{10}$	$\frac{1}{10}$	$\frac{1}{10}$	$\frac{1}{10}$	$\frac{1}{10}$	$\frac{1}{10}$	$\frac{1}{10}$	$\frac{1}{10}$	$\frac{1}{10}$

36 $\frac{2}{5} + \frac{1}{5} + \frac{1}{5} = $ _____

37 $\frac{4}{5} - \frac{1}{5} = $ _____

38 $1 - \frac{3}{10} = $ _____

39 $\frac{3}{10} + \frac{4}{10} + \frac{1}{10} = $ _____

40 $\frac{9}{10} - \frac{5}{10} = $ _____

41 $1 - \frac{3}{5} = $ _____

42 $\frac{2}{5} = \frac{}{10}$

43 $\frac{1}{2} = \frac{}{10}$

44 $\frac{3}{5} = \frac{}{10}$

45 $\frac{1}{5} = \frac{}{10}$

Write $<$, $=$ or $>$ in the circles to make each statement correct:

46 $\frac{1}{5} \bigcirc \frac{1}{2}$

47 $\frac{3}{5} \bigcirc \frac{3}{10}$

48 $\frac{4}{5} \bigcirc \frac{8}{10}$

49 $\frac{3}{5} \bigcirc \frac{1}{2}$

Here are four digits:  ① ③ ⑦ ④

Use each digit once only to form:

50 the largest possible four-digit number _____

51 the smallest possible four-digit number _____

Write the name of each shape:

52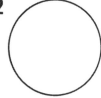

54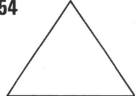

56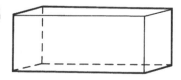

_____   _____   _____

53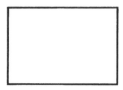

55

_____   _____

The graph shows how the 24 pupils of Class 2 come to school in the morning:

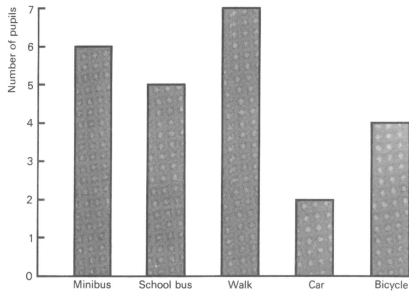

57 How many pupils come to school
by minibus?   _____

58 By which means do the most
pupils come to school?   _____

59 How many pupils do not walk to
school?   _____

60 How many pupils do not come to
school by the school bus?   _____

ANSWERS

1 NUMBER CONCEPTS

Exercise 1
1 73	**3** 14	**5** 38	**7** 40	**9** 712	**11** 78	**13** 305	**15** 400	**17** 228	**19** 905
2 63	**4** 90	**6** 111	**8** 534	**10** 840	**12** 45	**14** 620	**16** 62	**18** 110	**20** 625

Exercise 2
1 forty-three
2 nineteen
3 thirty-six
4 twenty-five
5 ninety-three
6 two hundred and thirteen
7 four hundred and eight
8 seventy-four
9 fifty
10 five hundred and ten
11 six hundred and forty-nine
12 nine hundred and seventy
13 three hundred and eight
14 two hundred and seventy-six
15 five hundred and forty-nine
16 six hundred and nineteen
17 four hundred and fourteen
18 seven hundred and seventy
19 two hundred and seventy-one
20 nine hundred and forty-three

Exercise 3
1 3000	**3** 2634	**5** 8049	**7** 7308	**9** 6011	**11** 7020	**13** 1000	**15** 8030	**17** 5005	**19** 4165
2 9000	**4** 3072	**6** 2084	**8** 9440	**10** 4545	**12** 4009	**14** 1006	**16** 9002	**18** 1811	**20** 5000

Exercise 4
1 three thousand, seven hundred and twenty-six
2 five thousand, one hundred and thirty-seven
3 six thousand, four hundred and three
4 two thousand, seven hundred and eight
5 three thousand
6 five thousand, six hundred and twenty
7 eight thousand and sixty-four
8 five thousand and seventy-six
9 two thousand and seven
10 nine thousand and one
11 three thousand, five hundred and seventy-six
12 four thousand and nine
13 three thousand and seventy-four
14 one thousand, six hundred and eight
15 nine thousand, four hundred and fifty-six
16 three thousand, four hundred and seventy
17 two thousand, seven hundred and nine
18 eight thousand and three
19 five thousand, six hundred
20 two thousand

Exercise 5
1 13	**2** 19	**3** 31	**4** 47	**5** 67	**6** 5	**7** 21	**8** 59	**9** 75	**10** 87

Exercise 6
1 23	**2** 29	**3** 47	**4** 13	**5** 39	**6** 73	**7** 47	**8** 19	**9** 53	**10** 61

Exercise 7
1 12	**2** 14	**3** 30	**4** 54	**5** 60	**6** 80	**7** 18	**8** 22	**9** 38	**10** 92

Exercise 8
1 14	**2** 22	**3** 38	**4** 6	**5** 8	**6** 28	**7** 36	**8** 52	**9** 70	**10** 102

Exercise 9
Odd numbers 7, 13, 23, 35, 71, 25, 15, 91, 95, 107, 125, 147
Even numbers 4, 6, 44, 12, 60, 102, 214, 316

Exercise 10
1 tens	**3** tens	**5** ones	**7** hundreds	**9** tens
2 ones	**4** hundreds	**6** tens	**8** hundreds	**10** ones

Exercise 11
1 hundreds	**4** thousands	**7** hundreds	**10** tens	**13** tens
2 ones	**5** thousands	**8** thousands	**11** hundreds	**14** thousands
3 hundreds	**6** tens	**9** ones	**12** hundreds	**15** ones

16 a 7   b 5   c 2   d 9
17 a tens   b thousands   c hundreds   d ones
18 a ones   b hundreds   c thousands   d tens

Exercise 12
1 <	**4** >	**7** <	**10** <	**13** >	**16** <	**19** >	**22** <	**25** >	**28** <
2 >	**5** >	**8** <	**11** <	**14** >	**17** <	**20** <	**23** >	**26** <	**29** >
3 <	**6** >	**9** >	**12** <	**15** <	**18** >	**21** >	**24** <	**27** <	**30** >

Exercise 13
1 19, 21, 29, 33, 37
2 11, 17, 22, 28, 30
3 28, 37, 39, 42, 45
4 78, 97, 118, 126, 134
5 86, 88, 97, 105, 112
6 168, 193, 203, 208, 215
7 881, 984, 997, 1048, 1076
8 1984, 2147, 2846, 3006, 3790
9 3857, 4719, 4962, 5120, 5204
10 5789, 5839, 5976, 6083, 6127

Exercise 14

1	23,	20,	19,	17,	11
2	46,	41,	37,	36,	29
3	28,	27,	24,	21,	19
4	101,	97,	84,	79,	76

5	110,	105,	95,	93,	87
6	154,	143,	139,	137,	126
7	387,	305,	296,	281,	257

8	1053,	1047,	894,	798,	763
9	1258,	1215,	1193,	1143,	1079
10	3105,	2963,	2851,	2742,	2543

Exercise 15

1 4 hundreds + 8 tens + 7 ones
2 2 hundreds + 9 tens + 1 one
3 5 hundreds + 3 tens + 6 ones
4 1 hundred + 0 tens + 4 ones
5 5 hundreds + 0 tens + 3 ones
6 6 hundreds + 7 tens + 0 ones
7 2 hundreds + 9 tens + 0 ones
8 2 thousands + 4 hundreds + 7 tens + 5 ones
9 1 thousand + 8 hundreds + 3 tens + 2 ones
10 3 thousands + 9 hundreds + 7 tens + 4 ones
11 6 thousands + 4 hundreds + 5 tens + 3 ones
12 8 thousands + 0 hundreds + 7 tens + 9 ones
13 5 thousands + 0 hundreds + 4 tens + 2 ones
14 3 thousands + 7 hundreds + 5 tens + 8 ones
15 6 thousands + 1 hundred + 4 tens + 7 ones
16 7 thousands + 5 hundreds + 3 tens + 0 ones
17 4 thousands + 6 hundreds + 0 tens + 1 one
18 9 thousands + 2 hundreds + 0 tens + 0 ones
19 8 thousands + 0 hundreds + 0 tens + 5 ones
20 9 thousands + 0 hundreds + 0 tens + 0 ones

Exercise 16

1 276	**3** 918	**5** 805	**7** 7473	**9** 6027	**11** 5006	**13** 3100	**15** 7505	**17** 5274	**19** 1305
2 450	**4** 503	**6** 2841	**8** 3952	**10** 1605	**12** 2500	**14** 8094	**16** 1006	**18** 6811	**20** 2580

Exercise 17

1 4 tens + 8 ones
2 2 tens + 6 ones
3 5 tens + 1 one
4 6 tens + 0 ones
5 8 tens + 0 ones
6 1 hundred + 7 tens + 5 ones
7 3 hundreds + 5 tens + 6 ones
8 7 hundreds + 6 tens + 0 ones
9 9 hundreds + 4 tens + 0 ones
10 4 hundreds + 0 tens + 1 one
11 2 hundreds + 0 tens + 3 ones
12 1 thousand + 3 hundreds + 5 tens + 8 ones
13 2 thousands + 7 hundreds + 4 tens + 6 ones
14 3 thousands + 0 hundreds + 7 tens + 1 one
15 8 thousands + 2 hundreds + 0 tens + 0 tens + 6 ones
16 5 thousands + 4 hundreds + 7 tens + 0 ones
17 7 thousands + 6 hundreds + 0 tens + 0 ones
18 6 thousands + 0 hundreds + 0 tens + 0 ones
19 4 thousands + 0 hundreds + 8 tens + 0 ones
20 6 thousands + 3 hundreds + 6 tens + 3 ones

Revision Exercise

1 a 780   b 3416   c 1900   d 6003   e 5020
2 a six hundred and four   b one thousand, four hundred and thirty-seven   c five thousand and forty-eight
   d six thousand and two   e eight thousand, three hundred
3 Odd numbers   15,   43,   61
   Even numbers   8,   14,   36,   28,   10,   72
4 a tens   b ones   c thousands   d hundreds   e tens
5 a >   b <   c <   d <   e >
6 a 9,   11,   16,   20,   27   b 8,   12,   13,   15,   19   c 14,   17,   18,   22,   26
   d 29,   33,   37,   41,   50   e 39,   47,   49,   58,   61
7 a 562   b 3457   c 1260   d 4038   e 7009
8 a 7 tens + 6 ones   b 2 hundreds + 8 tens + 4 ones
   c 3 thousands + 1 hundred + 5 tens + 7 ones   d 2 thousands + 6 hundreds + 0 tens + 8 ones
   e 1 thousand + 0 hundreds + 8 tens + 9 ones
9 a 3 hundreds + 2 tens + 5 ones   b 8 hundreds + 7 tens + 0 ones
   c 3 thousands + 5 hundreds + 2 tens + 6 ones   d 1 thousand + 2 hundreds + 0 tens + 8 ones
   e 6 thousands + 4 hundreds + 0 tens + 0 ones

2 OPERATIONS AND RELATIONS

Exercise 1

1 9	**5** 10	**9** 8	**13** 12	**17** 16	**21** 16	**25** 19	**29** 17
2 6	**6** 8	**10** 10	**14** 18	**18** 12	**22** 18	**26** 17	**30** 18
3 7	**7** 6	**11** 17	**15** 15	**19** 16	**23** 17	**27** 16	**31** 21
4 9	**8** 10	**12** 13	**16** 14	**20** 13	**24** 15	**28** 12	**32** 20

Exercise 2

1 69	**3** 79	**5** 79	**7** 797	**9** 767	**11** 697	**13** 889	**15** 799	**17** 7785	**19** 9792
2 79	**4** 89	**6** 868	**8** 668	**10** 869	**12** 889	**14** 859	**16** 5549	**18** 7768	**20** 7988

Exercise 3

1 60	**4** 90	**7** 84	**10** 76	**13** 88	**16** 61	**19** 79	**22** 90
2 81	**5** 82	**8** 85	**11** 73	**14** 96	**17** 73	**20** 79	**23** 90
3 63	**6** 75	**9** 93	**12** 85	**15** 78	**18** 70	**21** 73	**24** 90

Exercise 4

1 818	**4** 947	**7** 4984	**10** 7797	**13** 7685	**16** 5101	**19** 8122	**22** 8403	**25** 8206	**27** 9252
2 936	**5** 927	**8** 6885	**11** 3713	**14** 5662	**17** 6574	**20** 4414	**23** 5475	**26** 7024	**28** 6607
3 656	**6** 3681	**9** 8973	**12** 5745	**15** 6551	**18** 5442	**21** 6043	**24** 5401		

Exercise 5

1 4579	**3** 5467	**5** 4588	**7** 3487	**9** 6695	**11** 3811	**13** 3867	**15** 2808	**17** 5139	**19** 2468
2 2477	**4** 2777	**6** 2588	**8** 5889	**10** 7698	**12** 1878	**14** 1958	**16** 3202	**18** 6146	**20** 7287

Exercise 6

1 2, 4, 6, **8**, **10**, 12, **14**
2 1, 3, 5, **7**, **9**, 11, 13
3 15, 20, 25, **30**, 35, **40**, 45
4 22, 24, 26, **28**, 30, **32**, 34
5 1, 4, 7, 10, **13**, **16**, **19**

6 11, 21, **31**, 41, 51, **61**, **71**
7 8, 16, **24**, 32, 40, **48**, **56**
8 31, 33, 35, **37**, **39**, 41, **43**
9 4, 8, 12, 16, **20**, **24**, 28
10 1, 6, 11, 16, 21, **26**, 31

11 10, 20, 30, **40**, **50**, **60**, 70
12 60, 63, 66, **69**, **72**, **75**, 78
13 20, 40, 60, **80**, **100**, 120, **140**
14 10, 21, 32, 43, **54**, **65**, 76
15 **15**, 25, 35, 45, **55**, **65**, 75

Cross-number Puzzle for Addition

A 3	B 7	8	■	C 5	D 9
E 5	4	■	F 3	8	0
G 7	6	■	H 8	0	0
■	■	I 8	6	■	■
J 3	K 8	5	■	L 8	M 3
N 2	7	■	O 4	2	0

Exercise 7

1 991   2 1016   3 509   4 605   5 81   6 1645   7 660   8 1970   9 1050   10 240

Exercise 8

1 5	4 0	7 3	10 1	13 2	16 7	19 7	22 8	25 5	28 9
2 3	5 6	8 7	11 7	14 4	17 7	20 9	23 6	26 8	29 9
3 2	6 4	9 2	12 6	15 0	18 6	21 8	24 2	27 7	30 8

Exercise 9

1 43	3 62	5 55	7 50	9 53	11 143	13 143	15 115	17 193	19 245
2 64	4 18	6 23	8 22	10 62	12 115	14 151	16 118	18 223	20 453

Exercise 10

1 5	4 5	7 19	10 18	13 15	16 136	19 123	22 133
2 8	5 9	8 8	11 17	14 17	17 141	20 325	23 330
3 4	6 26	9 8	12 36	15 29	18 326	21 125	24 335

Exercise 11

1 2286	3 2458	5 1211	7 2533	9 2504	11 5151	13 3554	15 4152	17 5331	19 2142
2 1572	4 3507	6 4131	8 6037	10 6270	12 2313	14 5024	16 5101	18 1539	20 6303

Exercise 12

1 4116	4 2004	7 2737	10 4547	13 2655	16 2659	19 2876	22 978	24 487
2 2157	5 5053	8 2864	11 2777	14 4474	17 1367	20 3587	23 1799	25 93
3 1448	6 2484	9 4248	12 3455	15 2468	18 989	21 3697		

Exercise 13

1 45, 43, 41, **39**, **37**, **35**, 33
2 61, 56, 51, **46**, **41**, **36**, 31
3 40, 38, 36, 34, **32**, **30**, 28
4 **80**, 70, 60, 50, 40, 30, 20

5 **98**, 87, 76, 65, 54, **43**, 32
6 81, 80, 78, 75, **71**, **66**, 60
7 **87**, 82, 77, 72, 67, **62**, 57

8 36, 35, **34**, 33, 32, **31**, 30
9 50, 46, **42**, 38, 34, **30**, **26**
10 64, 54, 44, 34, **24**, 14, **4**

Cross-number Puzzle for Subtraction

A 3	B 3	■	C 3	D 5	E 1
F 2	2	G 2	■	H 6	4
■	■	I 8	0	■	■
J 1	K 8	3	■	L 4	M 4
N 5	8	■	O 6	0	4
P 1	3	5	■	Q 5	7

Exercise 14

1 138   2 266   3 66   4 86   5 103   6 239   7 60   8 283   9 367   10 365

Exercise 15

1 12	3 16	5 16	7 24	9 30	11 0	13 15	15 40	17 36	19 15
2 10	4 21	6 20	8 18	10 12	12 27	14 28	16 14	18 10	20 24

Exercise 16

1 86	4 305	7 115	10 372	13 496	16 486	19 688	22 790
2 99	5 328	8 195	11 486	14 580	17 344	20 876	23 828
3 208	6 64	9 156	12 645	15 832	18 900	21 530	24 552

Exercise 17

1 Multiply by 6.

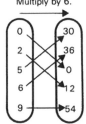

2 Multiply by 6.

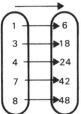

Exercise 18

1 246	**4** 540	**7** 3606	**10** 90	**13** 324	**16** 570	**19** 252	**22** 4854	**25** 1872	**27** 1440
2 360	**5** 420	**8** 4866	**11** 192	**14** 156	**17** 522	**20** 588	**23** 1830	**26** 1290	**28** 2010
3 486	**6** 1860	**9** 96	**12** 168	**15** 438	**18** 414	**21** 2442	**24** 3618		

Exercise 19

1 21	**2** 49	**3** 7	**4** 42	**5** 28	**6** 0	**7** 63	**8** 14	**9** 35	**10** 56

Exercise 20

1 287	**4** 350	**7** 707	**10** 4256	**13** 511	**16** 245	**19** 686	**22** 2450	**25** 2975	**27** 1988
2 420	**5** 630	**8** 4977	**11** 84	**14** 448	**17** 462	**20** 441	**23** 5684	**26** 2604	**28** 3759
3 567	**6** 770	**9** 1442	**12** 91	**15** 392	**18** 518	**21** 1680	**24** 1505		

Exercise 21

1 Multiply by 8.

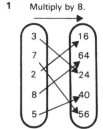

2 Multiply by 8.

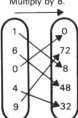

Exercise 22

1 568	**4** 480	**7** 520	**10** 688	**13** 776	**16** 2568	**19** 1672	**22** 3280	**25** 2208	**28** 7416
2 640	**5** 720	**8** 584	**11** 472	**14** 544	**17** 3600	**20** 4056	**23** 2504	**26** 2368	**29** 6672
3 408	**6** 336	**9** 232	**12** 672	**15** 616	**18** 2960	**21** 3288	**24** 2544	**27** 3480	**30** 3136

Exercise 23

1 18
2 63
3 36
4 0
5 81

6 Multiply by 9.

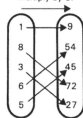

Exercise 24

1 639	**4** 459	**7** 6399	**10** 4500	**13** 2763	**16** 387	**19** 324	**22** 261	**25** 504	**28** 2574
2 720	**5** 810	**8** 5409	**11** 1440	**14** 3672	**17** 558	**20** 756	**23** 873	**26** 2223	**29** 7065
3 549	**6** 4860	**9** 8100	**12** 1260	**15** 4554	**18** 495	**21** 702	**24** 765	**27** 1782	**30** 5157

Exercise 25

1 30	**4** 80	**7** 310	**10** 160	**13** 650	**16** 760	**19** 8040	**22** 2250	**24** 4530
2 70	**5** 50	**8** 150	**11** 260	**14** 480	**17** 2700	**20** 3090	**23** 9460	**25** 8240
3 10	**6** 100	**9** 110	**12** 800	**15** 530	**18** 4600	**21** 5000		

Exercise 26

1 300	**3** 380	**5** 360	**7** 820	**9** 600	**11** 760	**13** 800	**15** 980
2 420	**4** 440	**6** 460	**8** 720	**10** 200	**12** 620	**14** 920	

Exercise 27

1 360	**3** 570	**5** 720	**7** 870	**9** 240	**11** 390	**13** 840	**15** 960
2 420	**4** 630	**6** 810	**8** 930	**10** 300	**12** 210	**14** 990	

Exercise 28

1 280	**3** 400	**5** 320	**7** 680	**9** 440	**11** 200	**13** 120	**15** 920
2 360	**4** 480	**6** 600	**8** 760	**10** 520	**12** 840	**14** 880	

Exercise 29

1 400, 2 300, 3 150, 4 250, 5 100, 6 200, 7 350, 8 800, 9 700, 10 650, 11 500, 12 450, 13 550, 14 750, 15 850

Exercise 30

1 180, 2 240, 3 250, 4 280, 5 480, 6 900, 7 780, 8 640, 9 850, 10 760, 11 510, 12 380, 13 700, 14 900, 15 800, 16 800, 17 950, 18 640, 19 720, 20 580, 21 750, 22 700, 23 690, 24 760, 25 3400, 26 3040, 27 1960, 28 2850, 29 2610, 30 4200

Exercise 31

1 360, 2 540, 3 560, 4 350, 5 480, 6 360, 7 720, 8 560, 9 1600, 10 2700, 11 3500, 12 4800, 13 4270, 14 2760, 15 6210, 16 6960, 17 3680, 18 6030, 19 2160, 20 4060, 21 6230, 22 2220, 23 8280, 24 7470, 25 4640, 26 2160, 27 5760, 28 5110, 29 3540, 30 3040

Cross-number Puzzle for Multiplication

A 3	B 2		C 1	D 3	E 8	
F 3	0	G 6		H 7	3	I 2
J 4	5	5		K 2	5	8
		L 7	M 6			7
N 9	O 6		P 5	Q 4	R 8	
S 3	4	T 4		U 2	3	V 8
W 6	8	8		X 4	7	5

Exercise 32

1 201, 2 720, 3 536, 4 455, 5 168, 6 75, 7 72, 8 600, 9 525, 10 882

Exercise 33

1 20, 2 10, 3 21, 4 23, 5 11, 6 142, 7 213, 8 100, 9 101, 10 10, 11 120, 12 101, 13 134, 14 201, 15 211

Exercise 34

1 23, 2 26, 3 13, 4 12, 5 12, 6 15, 7 19, 8 15, 9 16, 10 16, 11 252, 12 142, 13 62, 14 121, 15 120, 16 30, 17 263, 18 21, 19 193, 20 20, 21 122, 22 124, 23 484, 24 23

Exercise 35

1 276, 2 144, 3 176, 4 123, 5 135, 6 183, 7 123, 8 125, 9 145, 10 458, 11 179, 12 274, 13 142, 14 139, 15 367, 16 114, 17 132, 18 166, 19 154, 20 136, 21 257, 22 152, 23 188, 24 178, 25 137

Exercise 36

1 14 r1, 2 31 r2, 3 11 r3, 4 21 r2, 5 10 r4, 6 10 r3, 7 11 r3, 8 11 r2, 9 40 r1, 10 11 r1, 11 12 r1, 12 32 r1, 13 10 r4, 14 11 r3, 15 10 r1, 16 18 r1, 17 13 r1, 18 24 r1, 19 13 r3, 20 14 r1, 21 312 r2, 22 101 r2, 23 100 r5, 24 131 r2, 25 101 r1, 26 101 r1, 27 110 r3, 28 111 r3, 29 121 r1, 30 211 r2, 31 302 r1, 32 100 r4, 33 211 r3, 34 110 r1, 35 110 r7

Exercise 37

1 21 r1, 2 31 r3, 3 20 r4, 4 73 r1, 5 42 r2, 6 51 r1, 7 51 r1, 8 41 r3, 9 20 r3, 10 21 r2, 11 23 r2, 12 37 r1, 13 48 r1, 14 32 r1, 15 31 r1, 16 109 r3, 17 105 r3, 18 108 r1, 19 104 r6, 20 129 r2, 21 365 r1, 22 157 r1, 23 129 r3, 24 114 r1, 25 134 r3

Exercise 38

1 so 16 ÷ 2 = **8** and 16 ÷ 8 = **2**
2 so 28 ÷ 4 = **7** and 28 ÷ 7 = **4**
3 so 45 ÷ 9 = **5** and 45 ÷ 5 = **9**
4 so 42 ÷ 6 = **7** and 42 ÷ 7 = **6**
5 so 56 ÷ 8 = **7** and 56 ÷ 7 = **8**
6 so 72 ÷ 9 = **8** and 72 ÷ 8 = **9**
7 so 54 ÷ 9 = **6** and 54 ÷ 6 = **9**
8 so 30 ÷ 6 = **5** and 30 ÷ 5 = **6**
9 so 36 ÷ 4 = **9** and 36 ÷ 9 = **4**
10 so 40 ÷ 8 = **5** and 40 ÷ 5 = **8**
11 so 50 ÷ 5 = **10** and 50 ÷ 10 = **5**
12 so 70 ÷ 10 = **7** and 70 ÷ 7 = **10**

Exercise 39

1 9 × 3 = **27** so **27** ÷ 3 = 9 and **27** ÷ 9 = 3
2 6 × 8 = **48** so **48** ÷ 6 = 8 and **48** ÷ 8 = 6
3 7 × 5 = **35** so **35** ÷ 5 = 7 and **35** ÷ 7 = 5
4 7 × 9 = **63** so **63** ÷ 7 = 9 and **63** ÷ 9 = 7
5 4 × 8 = **32** so **32** ÷ 8 = 4 and **32** ÷ 4 = 8
6 4 × 7 = **28** so **28** ÷ 4 = 7 and **28** ÷ 7 = 4
7 8 × 8 = **64** so **64** ÷ 8 = 8
8 6 × 4 = **24** so **24** ÷ 4 = 6 and **24** ÷ 6 = 4
9 8 × 3 = **24** so **24** ÷ 3 = 8 and **24** ÷ 8 = 3
10 7 × 3 = **21** so **21** ÷ 3 = 7 and **21** ÷ 7 = 3

Exercise 40

1 8 × **2** = 16 so 16 ÷ 8 = **2** and 16 ÷ **2** = 8
2 3 × **4** = 12 so 12 ÷ 3 = **4** and 12 ÷ **4** = 3
3 2 × **9** = 18 so 18 ÷ 2 = **9** and 18 ÷ **9** = 2
4 10 × **5** = 50 so 50 ÷ 10 = **5** and 50 ÷ **5** = 10
5 3 × **9** = 27 so 27 ÷ 9 = **3** and 27 ÷ **3** = 9

6 $10 \times 6 = 60$
so $60 \div 6 = 10$
and $60 \div 10 = 6$

7 $9 \times 5 = 45$
so $45 \div 5 = 9$
and $45 \div 9 = 5$

8 $3 \times 8 = 24$
so $24 \div 8 = 3$
and $24 \div 3 = 8$

9 $8 \times 10 = 80$
so $80 \div 8 = 10$
and $80 \div 10 = 8$

10 $7 \times 6 = 42$
so $42 \div 6 = 7$
and $42 \div 7 = 6$

Cross-number Puzzle for Division

Exercise 41
1 26   2 30   3 29   4 45   5 36   6 28   7 25   8 26   9 23   10 24

Exercise 42
1 3   2 1
3 3   4 4
5 3   6 4
7 7   8 9
9 9   10 8
11 2   12 3
13 2   14 2
15 4   16 8
17 7   18 4
19 5   20 8

Exercise 43
1 $(7+3)+8 = 18$
2 $9+(5+5) = 19$
3 $(9+1)+7 = 17$
4 $5+(8+2) = 15$
5 $9+(9+1) = 19$
6 $8+(4+6) = 18$
7 $(19+1)+7 = 27$
8 $(17+3)+6 = 26$
9 $(18+2)+5 = 25$
10 $(2+28)+6 = 36$
11 $8+(39+1) = 48$
12 $(28+2)+30 = 60$
13 $(76+4)+9 = 89$
14 $(60+40)+17 = 117$
15 $(50+50)+8 = 108$
16 $(5 \times 2) \times 13 = 130$
17 $16 \times (5 \times 2) = 160$
18 $(5 \times 2) \times 27 = 270$
19 $(2 \times 50) \times 9 = 900$
20 $(4 \times 25) \times 8 = 800$
21 $(4 \times 25) \times 9 = 900$
22 $(5 \times 4) \times 25 = 500$
23 $(2 \times 2) \times 19 = 76$
24 $(3 \times 2) \times 26 = 156$
25 $(3 \times 2) \times 16 = 96$
26 $18 \times (6 \times 5) = 540$
27 $46 \times (8 \times 5) = 1840$
28 $(15 \times 2) \times 7 = 210$
29 $12 \times (15 \times 4) = 720$
30 $(8 \times 5) \times 17 = 680$

Exercise 44
1 0   4 0   7 11   10 14   13 19   16 0   19 0   22 0   25 0   28 47
2 7   5 0   8 0   11 18   14 0   17 31   20 26   23 18   26 43   29 34
3 9   6 5   9 0   12 36   15 28   18 26   21 15   24 63   27 0   30 0

Revision Exercise
1 a 69   b 975   c 96   d 163   e 594   f 877   g 601   h 761   i 5412   j 8346
2 a 354   b 44   c 2626   d 215   e 335   f 4624   g 3118   h 8062   i 3580   j 3646
3 a 86   b 116   c 326   d 17   e 488
4 a 249   b 3035   c 623   d 1296   e 6660   f 2864   g 870   h 3900   i 5840   j 4830
5 a 324   b 24   c 105   d 21   e 160   f 114   g 106 r5   h 428 r1   i 287 r1   j 178 r3
6 a 8   b 7   c 32   d 72   e 54
7 a 6   b 0   c 10   d 0   e 12   f 0   g 4   h 16   i 11   j 0
8 a 12   b 25   c 224   d 65   e 567
9 a 16   b 17   c 9   d 18   e 51

3 FRACTIONS

Exercise 1
1 $\frac{4}{8}$ 3 $\frac{6}{16}$ 5 $\frac{6}{16}$ 7 $\frac{10}{16}$ 9 $\frac{4}{8}$ 11 $\frac{12}{16}$ 13 $\frac{4}{16}$ 15 $\frac{8}{8}$
2 $\frac{8}{16}$ 4 $\frac{2}{16}$ 6 $\frac{12}{16}$ 8 $\frac{4}{16}$ 10 $\frac{8}{16}$ 12 $\frac{2}{8}$ 14 $\frac{4}{4}$

Exercise 2
1 $\frac{3}{4}$ 6 $\frac{1}{4}$ 11 $\frac{7}{8}$ 16 $\frac{5}{8}$ 21 $\frac{5}{16}$ 26 $\frac{2}{16}$ 31 $\frac{2}{16}$
2 $\frac{3}{4}$ 7 $\frac{3}{4}$ 12 $\frac{8}{8}$ 17 $\frac{1}{8}$ 22 $\frac{9}{16}$ 27 $\frac{5}{16}$ 32 $\frac{9}{16}$
3 $\frac{2}{4}$ 8 $\frac{3}{8}$ 13 $\frac{2}{8}$ 18 $\frac{7}{8}$ 23 $\frac{9}{16}$ 28 $\frac{2}{16}$ 33 $\frac{11}{16}$
4 $\frac{1}{4}$ 9 $\frac{4}{8}$ 14 $\frac{1}{8}$ 19 $\frac{3}{8}$ 24 $\frac{11}{16}$ 29 $\frac{4}{16}$ 34 $\frac{7}{16}$
5 $\frac{1}{4}$ 10 $\frac{7}{8}$ 15 $\frac{4}{8}$ 20 $\frac{3}{16}$ 25 $\frac{6}{16}$ 30 $\frac{5}{16}$ 35 $\frac{13}{16}$

Exercise 3
1 $\frac{2}{6}$ 2 $\frac{6}{12}$ 3 $\frac{8}{12}$ 4 $\frac{4}{12}$ 5 $\frac{8}{12}$ 6 $\frac{3}{3}$ 7 $\frac{6}{6}$ 8 $\frac{12}{12}$ 9 $\frac{10}{12}$

Exercise 4
1 $\frac{6}{10}$ 2 $\frac{8}{10}$ 3 $\frac{4}{10}$ 4 $\frac{5}{5}$ 5 $\frac{10}{10}$

Exercise 5
1 $\frac{2}{6}$ 5 $\frac{5}{6}$ 9 $\frac{2}{6}$ 13 $\frac{2}{12}$ 17 $\frac{8}{12}$ 21 $\frac{5}{12}$ 25 $\frac{1}{12}$ 29 $\frac{9}{12}$ 33 $\frac{7}{12}$
2 $\frac{3}{6}$ 6 $\frac{4}{6}$ 10 $\frac{1}{6}$ 14 $\frac{4}{12}$ 18 $\frac{10}{12}$ 22 $\frac{6}{12}$ 26 $\frac{4}{12}$ 30 $\frac{7}{12}$ 34 $\frac{12}{12}$
3 $\frac{4}{6}$ 7 $\frac{1}{6}$ 11 $\frac{5}{6}$ 15 $\frac{5}{12}$ 19 $\frac{9}{12}$ 23 $\frac{1}{12}$ 27 $\frac{6}{12}$ 31 $\frac{3}{12}$ 35 $\frac{11}{12}$
4 $\frac{5}{6}$ 8 $\frac{2}{6}$ 12 $\frac{2}{6}$ 16 $\frac{7}{12}$ 20 $\frac{11}{12}$ 24 $\frac{2}{12}$ 28 $\frac{5}{12}$ 32 $\frac{1}{12}$

Exercise 6

1 $\frac{2}{5}$	3 $\frac{3}{5}$	5 $\frac{4}{5}$	7 $\frac{1}{5}$	9 $\frac{1}{5}$	11 $\frac{2}{5}$	13 $\frac{3}{10}$	15 $\frac{7}{10}$	17 $\frac{1}{10}$	19 $\frac{4}{10}$	21 $\frac{3}{10}$
2 $\frac{3}{5}$	4 $\frac{4}{5}$	6 $\frac{5}{5}$	8 $\frac{1}{5}$	10 $\frac{2}{5}$	12 $\frac{2}{10}$	14 $\frac{5}{10}$	16 $\frac{9}{10}$	18 $\frac{4}{10}$	20 $\frac{5}{10}$	22 $\frac{6}{10}$

Exercise 7

1 < 4 > 7 > 10 < 13 < 16 < 19 = 22 > 25 > 28 =
2 > 5 < 8 < 11 < 14 > 17 > 20 > 23 < 26 > 29 =
3 > 6 > 9 = 12 = 15 = 18 > 21 = 24 = 27 > 30 >

Exercise 8

1 8   2 4   3 2   4 1   5 6   6 4   7 2   8 1   9 2   10 1

Exercise 9

1 Colour 4.
2 Colour 2.
3 Colour 4.
4 Colour 3.
5 Colour 3.
6 Colour any 3 parts.
7 Colour any 7 parts.
8 Colour any 2 parts.
9 Colour any 3 parts.
10 Colour any 5 parts.

Exercise 10

1 a $\frac{2}{5}$ b $\frac{3}{5}$
2 a $\frac{3}{8}$ b $\frac{5}{8}$
3 a $\frac{7}{8}$ b $\frac{1}{8}$
4 a $\frac{1}{4}$ b $\frac{3}{4}$
5 a $\frac{1}{2}$ b $\frac{1}{2}$
6 a $\frac{3}{5}$ b $\frac{2}{5}$
7 a $\frac{2}{3}$ b $\frac{1}{3}$
8 a $\frac{1}{6}$ b $\frac{5}{6}$
9 a $\frac{3}{8}$ b $\frac{5}{8}$
10 a $\frac{5}{12}$ b $\frac{7}{12}$

Exercise 11

1 6
2 9
3 12
4 30 minutes
5 50 grams
6 2
7 5
8 9
9 4 eggs
10 25 cents
11 2
12 5
13 8
14 15 seconds
15 25 cents
16 2
17 3
18 6
19 9 marbles
20 12 centimetres
21 2
22 4
23 5
24 10 apples
25 15 litres
26 1
27 2
28 5
29 9 metres
30 10 biscuits
31 1
32 3
33 4
34 7 kilograms
35 10 millilitres

Exercise 12

1 a $\frac{5}{8}$ b $\frac{3}{8}$
2 a $\frac{7}{10}$ b $\frac{3}{10}$
3 $\frac{2}{5}$
4 a $\frac{1}{3}$ b $\frac{2}{3}$
5 a $\frac{1}{4}$ b $\frac{3}{4}$
6 a $\frac{5}{8}$ b $\frac{3}{8}$
7 a $\frac{7}{10}$ b $\frac{3}{10}$
8 a $\frac{7}{10}$ b $\frac{3}{10}$
9 a $\frac{5}{6}$ b $\frac{1}{6}$
10 a $\frac{5}{12}$ b $\frac{7}{12}$

Revision Exercise

1 a $\frac{3}{6}$ b $\frac{2}{10}$ c $\frac{4}{8}$ d $\frac{14}{16}$ e $\frac{8}{16}$ f $\frac{3}{4}$ g $\frac{3}{5}$ h $\frac{8}{12}$ i $\frac{1}{3}$ j $\frac{2}{10}$

2 a $\frac{4}{5}$ b $\frac{3}{8}$ c $\frac{5}{8}$ d $\frac{7}{8}$ e $\frac{3}{12}$ f $\frac{9}{12}$ g $\frac{7}{12}$ h $\frac{5}{12}$ i $\frac{9}{12}$ j $\frac{9}{12}$
  k $\frac{3}{16}$ l $\frac{7}{16}$ m $\frac{7}{16}$ n $\frac{9}{16}$ o $\frac{13}{16}$ p $\frac{3}{6}$ q $\frac{5}{6}$ r $\frac{6}{6}$ or 1 s $\frac{11}{12}$ t $\frac{15}{16}$

3 a $\frac{1}{6}$ b $\frac{2}{6}$ c $\frac{3}{6}$ d $\frac{2}{5}$ e $\frac{1}{8}$ f $\frac{5}{8}$ g $\frac{2}{8}$ h $\frac{3}{8}$ i $\frac{1}{12}$ j $\frac{2}{12}$
  k $\frac{5}{12}$ l $\frac{3}{12}$ m $\frac{3}{16}$ n $\frac{5}{16}$ o $\frac{7}{16}$ p $\frac{4}{5}$ q $\frac{5}{6}$ r $\frac{5}{12}$ s $\frac{13}{16}$ t $\frac{3}{16}$

4 a 3 b 5 c 4 sweets d 20 minutes e 2
  f 3 g 8 h 5 i 2 litres j 16 marbles
  k 12 cents l 25 metres m 8 n 6 o 10
  p 6 grams q 2 litres r 5 minutes s 10 t 15

5 a < b > c = d > e < f < g < h > i < j =
  k = l > m < n = o > p < q < r < s = t >

6 a i $\frac{2}{5}$ ii $\frac{3}{5}$ b i $\frac{5}{12}$ ii $\frac{7}{12}$ c i $\frac{1}{5}$ ii $\frac{2}{5}$ iii $\frac{2}{5}$ d i $\frac{3}{6}$ ii $\frac{2}{6}$ iii $\frac{1}{6}$ e i $\frac{3}{8}$ ii $\frac{5}{8}$

4 MEASUREMENT AND GEOMETRY

Exercise 1

A 8 cm   B 5 cm   C 6 cm   D 9 cm   E 3 cm   F 7 cm   G 4 cm   H 10 cm   I 9 cm   J 2 cm

Exercise 2

1 10 cm 3 5 cm 5 35 cm 7 30 cm 9 40 cm 11 50 cm 13 60 cm 15 30 cm 17 70 cm 19 100 cm
2 20 cm 4 15 cm 6 25 cm 8 50 cm 10 30 cm 12 70 cm 14 65 cm 16 85 cm 18 95 cm 20 90 cm

Exercise 3

1 20 cm 4 40 cm 7 20 cm 10 80 cm 13 150 cm 16 90 cm 19 150 cm 22 160 cm 24 190 cm
2 10 cm 5 60 cm 8 30 cm 11 110 cm 14 180 cm 17 90 cm 20 140 cm 23 180 cm 25 200 cm
3 40 cm 6 50 cm 9 70 cm 12 110 cm 15 100 cm 18 100 cm 21 120 cm

Exercise 4

1 Mary
2 Ron
3 153 cm
4 Mr Rock
5 Crystal
6 Crystal
7 Michelle
8 Adrian
9 44 cm
10 79 cm

Exercise 5

1 a 500 mℓ b 250 mℓ   2 2   3 4   4 2

Exercise 6

1 C
2 D
3 125 mℓ
4 375 mℓ
5 A
6 E
7 B
8 B
9 3 times
10 4 times
11 2
12 125 mℓ
13 6
14 C
15 D

Exercise 7
1 B
2 F
3 A
4 D
5 F
6 A
7 D
8 B
9 7 kg
10 3 kg

Exercise 8
1 375 g
2 480 g
3 195 g
4 478 g
5 558 kg
6 597 kg
7 687 kg
8 528 kg
9 718 kg
10 969 kg
11 380 kg
12 573 kg
13 655 kg
14 530 kg
15 523 kg

Exercise 9
1 322 g
2 574 g
3 620 g
4 634 g
5 133 g
6 367 kg
7 361 kg
8 132 kg
9 145 kg
10 344 kg
11 323 kg
12 536 kg
13 456 kg
14 546 kg
15 374 kg

Exercise 10
1 240 g
2 375 g
3 450 g
4 210 g
5 900 g
6 600 g
7 530 g
8 915 g
9 875 g
10 876 g
11 861 kg
12 928 kg
13 945 kg
14 856 kg
15 625 kg
16 762 kg
17 990 kg
18 816 kg
19 960 kg
20 927 kg
21 858 kg
22 882 kg
23 725 kg
24 656 kg
25 825 kg

Exercise 11
1 200 g
2 230 g
3 320 g
4 400 g
5 230 g
6 120 g
7 310 g
8 300 g
9 210 g
10 120 g
11 200 g
12 220 g
13 110 g
14 100 g
15 110 g
16 110 g
17 275 g
18 393 g
19 250 g
20 282 g
21 130 g
22 162 g
23 167 g
24 144 g
25 122 g
26 134 g
27 155 g
28 132 g
29 123 g
30 124 g

Exercise 12
1 37 kg
2 10 kg
3 63 kg
4 700 g
5 225 g
6 71 kg
7 195 g
8 150 kg
9 375 g
10 a 19 kg   b 95 kg   c C   d D   e 12 kg

Exercise 13
1 1.30
2 3 o'clock
3 1.45
4 8.15

Exercise 14
1 2.20
2 12.05
3 6.25
4 10.10

Exercise 15
1 7.55
2 2.35
3 12.40
4 8.50

Exercise 16
1 2.30
2 6.55
3 5.40
4 11.45
5 10 o'clock
6 1.15
7 8 o'clock
8 8.35
9 11.20
10 5.10
11 8.25
12 1.50

180

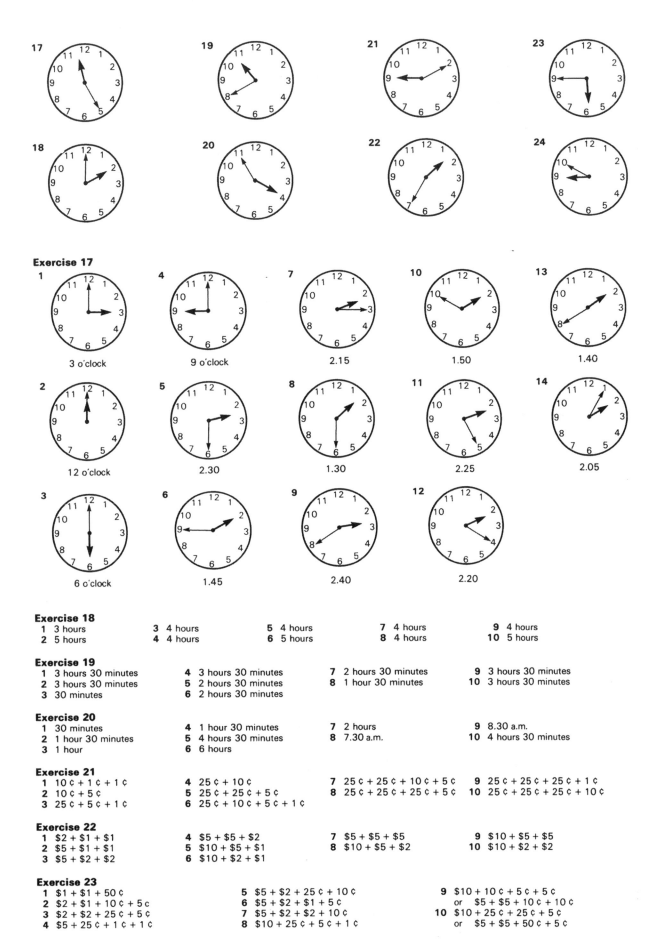

Exercise 17

1. 3 o'clock
4. 9 o'clock
7. 2.15
10. 1.50
13. 1.40

2. 12 o'clock
5. 2.30
8. 1.30
11. 2.25
14. 2.05

3. 6 o'clock
6. 1.45
9. 2.40
12. 2.20

Exercise 18
1. 3 hours
2. 5 hours
3. 4 hours
4. 4 hours
5. 4 hours
6. 5 hours
7. 4 hours
8. 4 hours
9. 4 hours
10. 5 hours

Exercise 19
1. 3 hours 30 minutes
2. 3 hours 30 minutes
3. 30 minutes
4. 3 hours 30 minutes
5. 2 hours 30 minutes
6. 2 hours 30 minutes
7. 2 hours 30 minutes
8. 1 hour 30 minutes
9. 3 hours 30 minutes
10. 3 hours 30 minutes

Exercise 20
1. 30 minutes
2. 1 hour 30 minutes
3. 1 hour
4. 1 hour 30 minutes
5. 4 hours 30 minutes
6. 6 hours
7. 2 hours
8. 7.30 a.m.
9. 8.30 a.m.
10. 4 hours 30 minutes

Exercise 21
1. 10 ¢ + 1 ¢ + 1 ¢
2. 10 ¢ + 5 ¢
3. 25 ¢ + 5 ¢ + 1 ¢
4. 25 ¢ + 10 ¢
5. 25 ¢ + 25 ¢ + 5 ¢
6. 25 ¢ + 10 ¢ + 5 ¢ + 1 ¢
7. 25 ¢ + 25 ¢ + 10 ¢ + 5 ¢
8. 25 ¢ + 25 ¢ + 25 ¢ + 5 ¢
9. 25 ¢ + 25 ¢ + 25 ¢ + 1 ¢
10. 25 ¢ + 25 ¢ + 25 ¢ + 10 ¢

Exercise 22
1. $2 + $1 + $1
2. $5 + $1 + $1
3. $5 + $2 + $2
4. $5 + $5 + $2
5. $10 + $5 + $1
6. $10 + $2 + $1
7. $5 + $5 + $5
8. $10 + $5 + $2
9. $10 + $5 + $5
10. $10 + $2 + $2

Exercise 23
1. $1 + $1 + 50 ¢
2. $2 + $1 + 10 ¢ + 5 c
3. $2 + $2 + 25 ¢ + 5 ¢
4. $5 + 25 ¢ + 1 ¢ + 1 ¢
5. $5 + $2 + 25 ¢ + 10 ¢
6. $5 + $2 + $1 + 5 ¢
7. $5 + $2 + $2 + 10 ¢
8. $10 + 25 ¢ + 5 ¢ + 1 ¢
9. $10 + 10 ¢ + 5 ¢ + 5 ¢
   or $5 + $5 + 10 ¢ + 10 ¢
10. $10 + 25 ¢ + 25 ¢ + 5 ¢
    or $5 + $5 + 50 ¢ + 5 ¢

181

Exercise 24

(Other answers are possible.)

	$10	$5	$2	$1	25 ¢	10 ¢	5 ¢	1 ¢
1			1	1	2			
2			2			1	1	
3		1		1		2	1	3
4		1			1	1	1	3
5		1	1		2			3
6		1	1	1	2	1		
7		1	1	1	3	2		
8		1	2		1		1	
9		1	2		3			2
10	1				3	1		1
11	1			1	2			1
12	1	1			2	2		4
13	1	1	2		1	2		2
14	2			1	1	1		3
15	2	1	2		3	1	1	3

Exercise 25

1 $7.26	3 $11.54	5 $9.23	7 $7.40
2 $13.25	4 $15.40	6 $13.32	8 $15.34

9 $14.21	11 $19.32	13 $30.46
10 $16.63	12 $23.45	14 $36.67

15 $41.55

Exercise 26

1 $2.15   3 $1.30   5 $6.00   7 $3.50   9 $2.90
2 $3.55   4 $2.55   6 $0.85 or 85 ¢   8 $3.05   10 a $5.60   b $4.40

Exercise 27

1 $3.97   3 $4.17   5 $7.42   7 $4.20   9 a $6.23   b $3.77
2 $6.10   4 $2.56   6 $5.67   8 $7.50   10 a $8.27   b $11.73

Exercise 28

1 square   3 triangle   5 triangle   7 square
2 circle   4 rectangle   6 rectangle   8 circle

Exercise 29

1 cone   2 sphere   3 cylinder   4 cube   5 cuboid

Exercise 30

Shape	Letter
Circle	H
Cone	F
Cube	D
Cuboid	B
Cylinder	C
Rectangle	I
Sphere	A
Square	G
Triangle	E

Exercise 31
1 3 faces   2 6 faces   3 6 faces

Revision Exercise
1 a A = 8 cm; B = 6 cm   b 15 cm   c 40 cm   d 30 cm   e 70 cm
2 a 597 g   b 363 g   c 635 g   d 824 g   e 156 g
3 a i 7.45   ii 1.20   iii 11.40   iv 10.05
   b i   ii   iii   iv

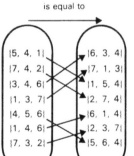

4 a 25 ¢ + 10 ¢ + 5 ¢   b 25 ¢ + 25 ¢ + 5 ¢ + 1 ¢   c 25 ¢ + 25 ¢ + 25 ¢ + 10 ¢   d 25 ¢ + 25 ¢ + 25 ¢ + 1 ¢
   e $1 + $1 + 25 ¢ + 25 ¢   f $10 + $2 + $1   g $5 + $2 + $2 + $2   h $5 + $5 + 25 ¢ + 10 ¢
   i $10 + $2 + $2 + 5 ¢   j $10 + $5 + 25 ¢ + 1 ¢
5 a Akari   b B   c 74 kg   d 1 hour
   e 8.15 a.m.   f $6.15   g $4.25   h 139 cm

5 SETS

Exercise 1
1 a Odd numbers {35, 21, 23, 5, 27}   b Even numbers {36, 6, 40, 28, 38}
2 a {kilometre, centimetre, metre}   b {gram, kilogram}   c {minute, hour}   d {litre, millilitre}
3 a {1 ¢, 5 ¢, 10 ¢, 25 ¢, $1.00}
   b {$1.00, $2.00, $5.00, $10.00, $20.00, $50.00, $100.00}
   c {Sunday, Monday, Tuesday, Wednesday, Thursday, Friday, Saturday}
   d {Jan., Feb., Mar., Apr., May, June, July, Aug., Sept., Oct., Nov., Dec.}
   e {11, 12, 13, 14, 15}
   f {13, 15, 17}
   g {22, 24, 26, 28}
   h {1, 2, 3, 4, 5, 6, 7, 8, 9, 10, 11, 12}
4 a odd numbers, 8   b even numbers, 9   c signs, 7   d shapes, 4   e rectangles, 5   f fractions, 6
   g solids or three-dimensional shapes, 5
5 a   is equal to   b   is equal to

a	
{4, 2, 6}	{4, 5, 6}
{1, 2, 3}	{2, 4, 6}
{5, 4, 6}	{1, 3, 2}
{9, 3, 5}	{3, 2, 4}
{2, 3, 4}	{5, 8, 6}
{6, 5, 8}	{3, 9, 5}

b	
{5, 4, 1}	{6, 3, 4}
{7, 4, 2}	{7, 1, 3}
{3, 4, 6}	{1, 5, 4}
{1, 3, 7}	{2, 7, 4}
{4, 5, 6}	{6, 1, 4}
{1, 4, 6}	{2, 3, 7}
{7, 3, 2}	{5, 6, 4}

6 GRAPHS

Exercise 1
1 6   2 pawpaw   3 mango   4 4   5 24

Exercise 2
1 the yellow house   2 14 points   3 the red house   4 the green house   5 the green house

Exercise 3
1 17 pupils   2 4 pupils   3 Tuesday   4 Monday and Friday

Exercise 4
1 6 pupils   2 grammar   3 social studies   4 31 pupils   5 5 pupils

Exercise 5
1 volleyball   2 basketball   3 football   4 2 pupils   5 lawn tennis   6 25 pupils

Exercise 6
1 7 hours   2 Wednesday   3 Monday   4 Tuesday   5 Thursday   6 45 hours

7 ASSESSMENT PAPERS (1-15)

Paper 1

1 13	11 25 cherries	21 12 lollipops	31 (+, +)
2 17	12 45 fish	22 39	32 (×, +), (×, −), or (÷, +)
3 14	13 69	23 15	33 310
4 16	14 368	24 19	34 2108
5 18	15 78	25 0	35 5016
6 6	16 580	26 19	36 6002
7 9	17 21	27 122 cm	37 4
8 3	18 23	28 (+, +) or (×, ×)	38 12
9 46	19 120	29 (+, ×)	39 5
10 20	20 292	30 (÷, +)	40 15

Paper 2

1 57	11 729	21 3.05	31 $\frac{4}{5}$
2 69	12 672	22 4.45	32 $\frac{3}{8}$
3 675	13 828	23 11.50	33 $\frac{5}{8}$
4 687	14 950	24 12 crayons	34 $\frac{1}{2}$
5 55 mangoes	15 96	25 $4.40	35 $\frac{1}{2}$
6 54	16 14	26 364	36 sphere
7 513	17 140	27 1865	37 rectangle
8 204	18 103	28 2506	38 circle
9 6431	19 135	29 4013	39 triangle
10 32 bubble-gum pictures	20 8 o'clock	30 $\frac{1}{5}$	40 square

Paper 3

1 776
2 6915
3 5036
4 858
5 19, 27, 30, 36, 43
6 12, 21, 28, 35, 41
7 37, 39, 42, 46, 50
8 38, 49, 53, 63, 71
9 33 pigs
10
11
12
13

14 tens
15 hundreds
16 thousands
17 ones
18 200 passengers
19 6
20 3
21 4
22 2
23 1
24 640
25 940
26 17
27 105
28 $2.45
29 $1.50
30 $3.25
31 $5.70
32 biscuits
33 9
34 4
35 24
36 7
37 14
38 4
39 3
40 8

Paper 4

1 7008	11 1740	21 $\frac{4}{8}$	31 24
2 342	12 2100	22 $\frac{12}{12}$ or 1	32 32
3 2501	13 2400	23 $\frac{7}{6}$	33 48
4 8040	14 2880	24 $\frac{6}{10}$	34 51 years old
5 5036	15 the yellow house	25 $\frac{2}{2}$ or 1	35 54 kg
6 4753	16 70 points	26 3 hours 30 minutes	36 2
7 2202	17 40 points	27 10 times	37 4
8 828	18 2 houses	28 12 members	38 5
9 306 r2	19 the blue house	29 6	39 11
10 810	20 $\frac{3}{5}$	30 10	40 4

Paper 5

1 2.30	11 20 cm	21 6	31 rectangle
2 10.25	12 10 cm	22 5	32 circle
3 9.10	13 40 cm	23 4	33 Colour 3.
4 5.40	14 45 cm	24 45 packets	34 Colour 2.
5 12.15 p.m.	15 50 cm	25 my father	35 Colour 2.
6 5802	16 840	26 $5.00	36 Colour 2.
7 3362	17 1040	27 square	37 7
8 1176	18 3000	28 cone	38 1
9 126	19 3560	29 triangle	39 0
10 15 cm	20 0	30 cube	40 5

Paper 6

1 15	7 6 cm	13 5 thousands + 7 hundreds + 2 tens + 3 ones
2 17	8 9 cm	14 6 thousands + 8 hundreds + 0 tens + 0 ones
3 21	9 4 cm	15 45 bunches
4 13	10 1945 limes	16 14
5 39	11 2 hundreds + 5 tens + 7 ones	17 781
6 7 cm	12 3 thousands + 4 hundreds + 0 tens + 6 ones	18 5264

184

19 856	27 3 times
20 125	28 D
21 3050	29 250 mℓ
22 1450	30 (+, −)
23 1900	31 (+, ×)
24 3700	32 (×, −)
25 D	33 (×, +), (×, −), (÷, +), or (÷, −)
26 A	

34 (÷, ×)	
35 (−, ×)	
36 four thousand, six hundred and fifteen	
37 one thousand and seventy-three	
38 three thousand and seven	
39 four hundred and eighty	
40 five thousand, six hundred	

Paper 7

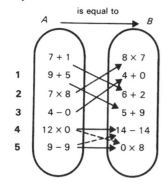

6 10 cm	
7 20 cm	
8 40 cm	
9 40 cm	
10 60 cm	
11 30 cm	
12 4768	
13 6625	
14 1336	
15 125	
16 19 green marbles	
17 $5.50	

18 490	
19 1820	
20 2730	
21 17	
22 Shekira	
23 Keisha	
24 Isha and Vena	
25 Shekira	
26 Ayesha and Keisha	
27 Isha	
28 Vena's	
29 9	

30 7	
31 13	
32 8	
33 8	
34 27	
35 5	
36 4	
37 9th October	
38 23rd October	
39 9732	
40 2379	

Paper 8

1 $\frac{7}{8}$	
2 $\frac{5}{8}$	
3 $\frac{2}{8}$	
4 $\frac{6}{4}$	
5 $\frac{1}{4}$	
6 $\frac{3}{8}$	
7 $\frac{1}{8}$	
8 $\frac{1}{2}$	
9 $\frac{5}{8}$	
10 $\frac{3}{8}$	

11 2351	
12 5163	
13 1709	
14 6045	
15 12 boys	
16 720	
17 1440	
18 2160	
19 3680	
20 15 apples	

21 Akari	
22 Carlo	
23 9 kg	
24 Carlo	
25 74 kg	
26 7814	
27 3454	
28 5481	
29 104	
30 24	

31 17	
32 56	
33 6	
34 tens	
35 hundreds	
36 thousands	
37 ones	
38 $5.20	
39 $4.80	
40 6 faces	

Paper 9

1 2562	
2 7200	
3 187	
4 3009	
5 5010	
6 5840	
7 6058	
8 3840	
9 105	
10 3 hours	

11 1 hour	
12 2 hours	
13 2 hours 30 minutes	
14 9th March	
15 23rd March	
16 640	
17 2160	
18 2880	
19 4720	
20 17 kg	

21 Corncurls	
22 sweet biscuits	
23 popcorn	
24 4 pupils	
25 Cheesezees	
26 4 pupils	
27 30 pupils	
28 9	
29 24	
30 19	

31 8	
32 15	
33 12	
34 8	
35 6	
36 4	
37 3	
38 2	
39 $\frac{7}{12}$	
40 $\frac{5}{12}$	

Paper 10

1 23	
2 24	
3 7	
4 24	
5 796	
6 5204	
7 1668	
8 134	
9 7	
10 2	

11 **2** thousands + **1** hundred + **7** tens + **6** ones

12 **3** thousands + **0** hundreds + **5** tens + **4** ones

13 **5** thousands + **0** hundreds + **0** tens + **3** ones

14 **2** hundreds + **9** tens + **7** ones

15 $\frac{2}{4}$	
16 $\frac{2}{6}$	
17 $\frac{6}{12}$	
18 $\frac{3}{12}$	
19 $\frac{4}{6}$	
20 $\frac{8}{12}$	
21 $\frac{10}{12}$	
22 $\frac{9}{12}$	
23 a square	

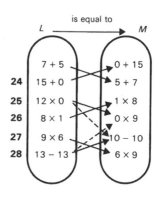

29	7 o'clock	32	10.30	35	(×, ×)	38	(+, ×)
30	11.10	33	1.45 p.m.	36	(÷, +)	39	(−, ×)
31	2.40	34	(×, +)	37	(−, −)	40	3 faces

Paper 11

1	987	9	32, even	17	4	25	7	33	10 m
2	6211	10	41, odd	18	the yellow house	26	9	34	1457
3	5122	11	40, even	19	the red house	27	4	35	7541
4	5054	12	57, odd	20	the red house	28	6	36	649
5	570	13	62, even	21	the green house	29	2	37	63
6	1920	14	8	22	22 medals	30	the library	38	85
7	106	15	6	23	the yellow house	31	605 m	39	108
8	72	16	5	24	8	32	535 m	40	1 ℓ container

Paper 12

1	98	9	48	17	$\frac{3}{10}$	25	18	33	7
2	578	10	74	18	368	26	105	34–9	357,
3	7683	11	50	19	95	27	23 elephants		375,
4	103	12	326	20	2863	28	1		537,
5	5014	13	6291	21	1896	29	0		573,
6	7020	14	$\frac{5}{12}$	22	24 pupils	30	0		735,
7	3108	15	$\frac{7}{12}$	23	15	31	5		753 (any order)
8	2037	16	$\frac{7}{10}$	24	114	32	0	40	$1.75

Paper 13

1	four thousand, three hundred and sixteen	15	870	29	<
2	two thousand and seventy-six	16	1380	30	=
3	one thousand	17	960	31	>
4	five thousand, three hundred	18	1610	32	$\frac{3}{4}$
5	7.30	19	1475 flying fish	33	$\frac{2}{8}$ or $\frac{1}{4}$
6	2 o'clock	20	15 fish cakes	34	1
7	11.15	21	the reading book	35	$\frac{5}{8}$
8	2.35	22	the colouring book	36	$\frac{3}{4}$
9	7570	23	the reading book and the colouring book	37	$\frac{1}{2}$
10	3907	24	$9.95	38	2 guavas
11	2590	25	$10.05	39	8 guavas
12	72	26	>	40	72 seeds
13	Carl	27	=		
14	6 cm	28	>		

Paper 14

1	18	9	2 children	17	35 cm	25	2720	33	$0.60 or 60 ¢
2	33	10	Friday	18	79	26	4600	34	87
3	41	11	105 pupils	19	106	27	3690	35	6
4	25	12	20 cm	20	4856	28	3529	36	0
5	25 sweets	13	20 cm	21	5460	29	5184	37	7
6	781 limes	14	45 cm	22	7 children	30	7080	38	4
7	20 children	15	40 cm	23	4 plums	31	6003	39	0
8	Tuesday	16	50 cm	24	1110	32	$3.15	40	12

Paper 15

1	1722	9	1 hour, 30 minutes	17	$5.95	25	15	33	16
2	178 r1	10	7542	18	$6.70	26	8	34	29
3	4506	11	2457	19	sphere	27	48	35	3
4	3880	12	André	20	square	28	5	36	7
5	11.30	13	4 kg	21	cuboid	29	5	37	9
6	7.45	14	Ayanna	22	cube	30	7	38	5
7	10.00	15	Shakira	23	rectangle	31	5	39	$\frac{5}{8}$
8	2.20	16	$1.25	24	circle	32	19	40	$\frac{3}{8}$

8 TWO TESTS

Test 1

1	94	9	104	17	12	25	>
2	5010	10	4184	18	8	26	=
3	2006	11	830	19	36	27	=
4	3084	12	5820	20	5	28	>
5	61	13	482	21	6	29	the bag of potatoes
6	385	14	75	22	8	30	12 bags
7	316	15	15 friends	23	8	31	98
8	105	16	6 faces	24	113	32	17

33 16	**40** $2.55		**47** 8023		**54** 2 o'clock	
34 4	**41** 163 cm		**48** 4407		**55** 7.15	
35 21	**42** four hundred and thirty-six		**49** 5		**56** 9.10	
36 9	**43** one thousand, seven hundred and eight		**50** 3		**57** $\frac{3}{10}$	
37 16	**44** five thousand and six		**51** 2		**58** $\frac{7}{10}$	
38 8	**45** two thousand and twenty-seven		**52** 0		**59** 5	
39 8	**46** 3586		**53** 11.40		**60** 2	

Test 2

1 185
2 4256
3 6073
4 3440
5 4 **hundreds** + 7 **tens** + 2 **ones**
6 2 **thousands** + 0 **hundreds** + 5 **tens** + 6 **ones**
7 7 **thousands** + 3 **hundreds** + 8 **tens** + 0 **ones**
8 $4.05
9 $5.95
10 4
11 32
12 Pedro
13 Pedro
14 Ron
15 Ron
16 André

17

18

19

20

21 10.45 a.m.
22 23
23 hundreds
24 ones
25 thousands
26 tens
27 30 cm
28 40 cm
29 50 cm
30 40 cm
31 53
32 8 **hundreds** + 7 **tens** + 3 **ones**
33 4 **thousands** + 0 **hundreds** + 8 **tens** + 2 **ones**
34 54 years old

35 $5.72
36 $\frac{4}{5}$
37 $\frac{3}{5}$
38 $\frac{7}{10}$
39 $\frac{8}{10}$
40 $\frac{4}{10}$
41 $\frac{2}{5}$
42 $\frac{4}{10}$
43 $\frac{5}{10}$
44 $\frac{6}{10}$
45 $\frac{2}{10}$
46 <
47 >
48 =
49 >
50 7431
51 1347
52 circle
53 rectangle
54 triangle
55 cone
56 cuboid
57 6 pupils
58 walk
59 17 pupils
60 19 pupils

UNIVERSITY PRESS

Great Clarendon Street, Oxford, OX2 6DP, United Kingdom

Oxford University Press is a department of the University of Oxford.
It furthers the University's objective of excellence in research, scholarship,
and education by publishing worldwide. Oxford is a registered trade mark of
Oxford University Press in the UK and in certain other countries

First published by Nelson Thornes Ltd in 1989
This edition published by Oxford University Press in 2014

British Library Cataloguing in Publication Data
Data available

978-0-7487-0110-0

40 39 38 37

Printed in India by Gopsons Papers Ltd., Sivakasi

Acknowledgements

Page make-up: Tech-Set

Although we have made every effort to trace and contact all
copyright holders before publication this has not been possible in all
cases. If notified, the publisher will rectify any errors or omissions at
the earliest opportunity.